Advances in Intelligent Systems and Computing

Volume 1139

AF846912

Series Editor

Janusz Kacprzyk, Systems Research Institute, Polish Academy of Sciences, Warsaw, Poland

Advisory Editors

Nikhil R. Pal, Indian Statistical Institute, Kolkata, India

Rafael Bello Perez, Faculty of Mathematics, Physics and Computing, Universidad Central de Las Villas, Santa Clara, Cuba

Emilio S. Corchado, University of Salamanca, Salamanca, Spain

Hani Hagras, School of Computer Science and Electronic Engineering, University of Essex, Colchester, UK

László T. Kóczy, Department of Automation, Széchenyi István University, Gyor, Hungary

Vladik Kreinovich, Department of Computer Science, University of Texas at El Paso, El Paso, TX, USA

Chin-Teng Lin, Department of Electrical Engineering, National Chiao Tung University, Hsinchu, Taiwan

Jie Lu, Faculty of Engineering and Information Technology, University of Technology Sydney, Sydney, NSW, Australia

Patricia Melin, Graduate Program of Computer Science, Tijuana Institute of Technology, Tijuana, Mexico

Nadia Nedjah, Department of Electronics Engineering, University of Rio de Janeiro, Rio de Janeiro, Brazil

Ngoc Thanh Nguyen, Faculty of Computer Science and Management, Wrocław University of Technology, Wrocław, Poland

Jun Wang, Department of Mechanical and Automation Engineering, The Chinese University of Hong Kong, Shatin, Hong Kong

The series “Advances in Intelligent Systems and Computing” contains publications on theory, applications, and design methods of Intelligent Systems and Intelligent Computing. Virtually all disciplines such as engineering, natural sciences, computer and information science, ICT, economics, business, e-commerce, environment, healthcare, life science are covered. The list of topics spans all the areas of modern intelligent systems and computing such as: computational intelligence, soft computing including neural networks, fuzzy systems, evolutionary computing and the fusion of these paradigms, social intelligence, ambient intelligence, computational neuroscience, artificial life, virtual worlds and society, cognitive science and systems, Perception and Vision, DNA and immune based systems, self-organizing and adaptive systems, e-Learning and teaching, human-centered and human-centric computing, recommender systems, intelligent control, robotics and mechatronics including human-machine teaming, knowledge-based paradigms, learning paradigms, machine ethics, intelligent data analysis, knowledge management, intelligent agents, intelligent decision making and support, intelligent network security, trust management, interactive entertainment, Web intelligence and multimedia.

The publications within “Advances in Intelligent Systems and Computing” are primarily proceedings of important conferences, symposia and congresses. They cover significant recent developments in the field, both of a foundational and applicable character. An important characteristic feature of the series is the short publication time and world-wide distribution. This permits a rapid and broad dissemination of research results.

**** Indexing: The books of this series are submitted to ISI Proceedings, EI-Compendex, DBLP, SCOPUS, Google Scholar and Springerlink ****

More information about this series at http://www.springer.com/series/11156

Atulya K. Nagar · Kusum Deep ·
Jagdish Chand Bansal · Kedar Nath Das
Editors

Soft Computing for Problem Solving 2019

Proceedings of SocProS 2019, Volume 2

Springer

Editors
Atulya K. Nagar
School of Mathematics,
Computer Science and Engineering
Liverpool Hope University
Liverpool, UK

Jagdish Chand Bansal
Department of Mathematics
South Asian University
New Delhi, Delhi, India

Kusum Deep
Department of Mathematics
Indian Institute of Technology Roorkee
Roorkee, India

Kedar Nath Das
Department of Mathematics
National Institute of Technology Silchar
Silchar, India

ISSN 2194-5357 ISSN 2194-5365 (electronic)
Advances in Intelligent Systems and Computing
ISBN 978-981-15-3286-3 ISBN 978-981-15-3287-0 (eBook)
https://doi.org/10.1007/978-981-15-3287-0

© Springer Nature Singapore Pte Ltd. 2020
This work is subject to copyright. All rights are reserved by the Publisher, whether the whole or part of the material is concerned, specifically the rights of translation, reprinting, reuse of illustrations, recitation, broadcasting, reproduction on microfilms or in any other physical way, and transmission or information storage and retrieval, electronic adaptation, computer software, or by similar or dissimilar methodology now known or hereafter developed.
The use of general descriptive names, registered names, trademarks, service marks, etc. in this publication does not imply, even in the absence of a specific statement, that such names are exempt from the relevant protective laws and regulations and therefore free for general use.
The publisher, the authors and the editors are safe to assume that the advice and information in this book are believed to be true and accurate at the date of publication. Neither the publisher nor the authors or the editors give a warranty, expressed or implied, with respect to the material contained herein or for any errors or omissions that may have been made. The publisher remains neutral with regard to jurisdictional claims in published maps and institutional affiliations.

This Springer imprint is published by the registered company Springer Nature Singapore Pte Ltd.
The registered company address is: 152 Beach Road, #21-01/04 Gateway East, Singapore 189721, Singapore

Preface

We are delighted that the 9th International Conference on Soft Computing for Problem Solving, SocProS 2019, took place at Liverpool Hope University, UK, during 02–04 September 2019. This was a particularly important event as it coincided with the 175th anniversary of the foundation of Liverpool Hope University (LHU). The SocProS conference series has a glorious history; the earlier editions of the conference have been organised in various prestigious institutions of India. For the very first time, this event in the series of the conference was hosted outside of India at Liverpool in the UK. This is significant because along with IIT Roorkee and South Asian University in India, LHU has been one of the key institutions that initiated this prestigious meeting. Continuing the trend, once again the 9th edition of this conference touched many milestones in terms of quality research papers and fruitful discussions. The theme of SocProS 2019 was "*Unlocking the Power and Impact of Artificial Intelligence*".

This proceedings as an outcome of the 9th meeting of the SocProS community includes a collection of selected high-quality articles on various topics related to soft computing and artificial intelligence and their applications. The book is being prepared in two volumes to cover the recent advances and challenges in the themes of machine learning, neural networks, scientific computing, and intelligent systems and includes several chapters addressing the problems arising in real-life applications comprising that of image classification, deep learning, fuzzy systems, flow shop scheduling, support vector machines, mobile robot path planning, P-systems, machine learning, and spiking neural networks, to name a few contributions. We have also tried to capture the impact aspects of research in this area, particularly impact beyond the academic world. We have made further efforts in this direction to embed impact as part of our conference series, and going forward we very much hope that, as was agreed at the conference, we will continue to mainstream impact in our work and intensify our efforts to reach out to non-academic beneficiaries and users to realise the impact from our research.

Highlighting theoretical perspectives and empirical research, it is hoped that this edited volume will prove to be a comprehensive reference source for researchers, practitioners, students, and professionals interested in the current advancements and efficient use of soft computing as well as in making the impact happen. We express our heartfelt gratitude to all the authors, reviewers, and Springer personnel for their motivation and patience.

Liverpool, UK
September 2019

Atulya K. Nagar
Kusum Deep
Jagdish Chand Bansal
Kedar Nath Das

Contents

Exponential Adaptive Strategy in Spider Monkey Optimization Algorithm 1
Apoorva Sharma, Nirmala Sharma, Harish Sharma, and Jagdish Chand Bansal

Development of Fuzzy Knowledge-Based System for Water Quality Assessment in River Ganga 17
Praveen Kumar Shukla

A Hybrid Framework for Fire Outbreak Detection Based on Interval Type-2 Fuzzy Logic and Flower Pollination Algorithm 27
Uduak A. Umoh, Udoinyang G. Inyang, and Emmanuel E. Nyoho

Using Convolutional Neural Networks to Predict Colon Cancer Patients Survival 47
Rawan Gedeon, Atulya K. Nagar, and Raouf Naguib

An Array P System Based on a Variant of Pure 2D Context-Free Grammars 57
P. S. Azeezunnisha, S. Hemalatha, Sastha Sriram, and Atulya K. Nagar

Predictions of Weekly Slope Movements Using Moving-Average and Neural Network Methods: A Case Study in Chamoli, India 67
Praveen Kumar, Priyanka, Ankush Pathania, Shubham Agarwal, Naresh Mali, Ravinder Singh, Pratik Chaturvedi, K. V. Uday, and Varun Dutt

Two-Stage History Matching for Hydrology Models via Machine Learning 83
Dewi Tjia, Ritu Gupta, and Muhammad Alam

An Intelligent System for Diagnosis of Diabetic Retinopathy 97
Saroj Kr. Biswas, Rohit Upadhya, Nipan Das, Dolly Das, Manomita Chakraborty, and Biswajit Purkayastha

Markov Chain Models for the Near Real-Time Forecasting of Australian Football League Match Outcomes . . . 111
Casey Josman, Ritu Gupta, and Sam Robertson

Genetically Optimized Deep Neural Learning for Breast Cancer Prediction . . . 127
Suchitra Agrawal, Aruna Tiwari, and Ishan Goel

Optimization of Lycopene Extraction from Tomato Processing Waste Skin Using Harmony Search Algorithm . . . 141
Assif Assad, Kusum Deep, Neil Buckley, and Atulya K. Nagar

Meme-Based Computational Optimization Framework . . . 155
Felis Dwiyasa, Meng-Hiot Lim, Ren-Xiang Foo, and Shi-Wei Jason Teo

Heterogeneous Multi-robot Mission Planning for Coordinated Tasks Execution . . . 167
Felis Dwiyasa, Meng-Hiot Lim, Pyo Kang, Ren-Xiang Foo, and Shi-Wei Jason Teo

Development of Cost-Effective Endurance Test Rig with Integrated Algorithm for Safety . . . 175
Emanuele Lindo Secco, Rashid Abdulrahman, Ian Felmeri, and Atulya K. Nagar

Development of an Algorithm for the EMG Control of Prosthetic Hand . . . 191
Emanuele Lindo Secco, Philippe Caddet, and Atulya K. Nagar

Improving Data Quality in the Cargo Industry with Modern Recurrent Neural Network Architecture . . . 199
Lewis Wong, Declan O'Connor, Neil Buckley, and Atulya K. Nagar

Temporal Convolution in Spiking Neural Networks: A Bio-mimetic Paradigm . . . 211
David Reid and Emanuele Lindo Secco

Author Index . . . 223

About the Editors

Atulya K. Nagar holds the Foundation Chair as Professor of Mathematical Sciences, and is the Pro-Vice-Chancellor for Research and Dean of the Faculty of Science at Liverpool Hope University, United Kingdom. He is also the Head of the School of Mathematics, Computer Science and Engineering, which he established at the University. He is an internationally respected scholar working at the cutting edge of theoretical computer science, applied mathematical analysis, operations research, and systems engineering. He received a prestigious Commonwealth Fellowship to pursue his doctorate (DPhil) in Applied Nonlinear Mathematics, which he earned from the University of York (UK) in 1996. He holds a BSc (Hons), an MSc and MPhil (with distinction) in Mathematical Physics from the MDS University of Ajmer, India. His research expertise spans both applied mathematics and computational methods for nonlinear, complex, and intractable problems arising in science, engineering and industry.

Prof. Kusum Deep is a Professor at the Department of Mathematics, Indian Institute of Technology Roorkee. Her research interests include numerical optimization, nature inspired optimization, computational intelligence, genetic algorithms, parallel genetic algorithms, and parallel particle swarm optimization.

Dr. Jagdish Chand Bansal is an Assistant Professor at the South Asian University, New Delhi, India and visiting research fellow at Liverpool Hope University, UK. He has an excellent academic record and is a leading researcher in the field of swarm intelligence. He has published numerous research papers in respected international and national journals.

Dr. Kedar Nath Das is an Assistant Professor at the Department of Mathematics, National Institute of Technology, Silchar, Assam, India. Over the past 10 years, he has made substantial contributions to research on soft computing, and has published several papers in prominent national and international journals. His chief area of interest is evolutionary and bio-inspired algorithms for optimization.

About the Editors

Exponential Adaptive Strategy in Spider Monkey Optimization Algorithm

Apoorva Sharma, Nirmala Sharma, Harish Sharma, and Jagdish Chand Bansal

Abstract Spider monkey optimization algorithm (SMOA) is one of the powerful techniques in the arena of swarm intelligence (SI)-based strategies. This article proposes a modified variant of SMOA that is based on an exponential adaptive strategy for step size. During the search of the optimal solution, this exponential strategy is used to adjust the step size so that it can speed up the convergence ability of the swarm. The proposed algorithm is termed as exponential adaptive spider monkey optimization (EASMO) algorithm. This evinced algorithm is tested over 14 standard optimization problems to examine its authenticity. Further, the obtained results are compared with the artificial bee colony (ABC), differential evolution (DE), Gbest-guided artificial bee colony (GABC), particle swarm optimization (PSO), SMOA, and three of its momentous variants, namely levy flight SMOA (LFSMOA), modified limacon SMOA (MLSMOA), and power law-based local search in SMOA (PLSMOA). The analysis of the results proved the competence of EASMO in the field of SI-based strategies.

Keywords Population-based optimization · Spider monkey optimization algorithm · Swarm intelligence-based algorithms · Exponential adaptive strategy

A real one.

A. Sharma (✉) · N. Sharma · H. Sharma
Rajasthan Technical University, Kota, Rajasthan, India
e-mail: apoorvasharma2628@gmail.com

N. Sharma
e-mail: nsharma@rtu.ac.in

H. Sharma
e-mail: harish.sharma0107@gmail.com

J. Chand Bansal
South Asian University, New Delhi, India
e-mail: jcbansal@gmail.com

© Springer Nature Singapore Pte Ltd. 2020

A. K. Nagar et al. (eds.), *Soft Computing for Problem Solving 2019*,
Advances in Intelligent Systems and Computing 1139,
https://doi.org/10.1007/978-981-15-3287-0_1

1 Introduction

Researchers have analyzed the behavior of several systems in nature that is able to cope efficiently with many optimization problems (OPs), and these systems could be used to resolve various complex optimization queries [2]. Population-based optimization algorithms are used to find near-optimal solutions for difficult OPs [3]. Evolutionary algorithms (EAs) and swarm intelligence (SI)-based algorithms are two essential categories of population-based optimization algorithms. EAs use mechanisms motivated by biological evolution, like selection, crossover, mutation, etc. [4]. SI is a metaheuristic method in the field of artificial intelligence (AI). To solve actual world OPs, SI-dependent algorithms are used because of its significant potential [5, 11, 12, 20–23]. Ant colony optimization (ACO) [2], particle swarm optimization (PSO) [8], bacterial foraging optimization (BFO) [10], artificial bee colony (ABC) optimization [7], etc. are some significant SI-based algorithms. Spider monkey optimization algorithm (SMOA) also come under the category of SI-based algorithms. Food searching behavior of spider monkeys is used as a basis for developing SMOA. With the ability to solve numerical OPs, it also has a inherit drawback like other population-based optimization algorithms such as converges to the local optima probabilistically and sometimes get stuck at local optima [1].

To maintain an appropriate consistency between exploration and exploitation, it is necessary for the algorithm to first use the variation process for few iterations and then use the selection process for the rest of the iterations. In order to reduce the chances to jump the optimal solution and enhance the exploitation proficiency of the SMOA, an exponential adaptive strategy is applied with SMOA. This adaptive strategy decreases the step size exponentially. The propound algorithm is titled as an exponential adaptive spider monkey optimization (EASMO) algorithm. The authenticity of the proposed EASMO is evaluated using 14 benchmark optimization functions. The obtained outcomes are analyzed and compared with eight state-of-the-art algorithms. The obtained results validate the proposed approach.

The arrangement of the remaining article is as follows: Sect. 2 discusses SMOA. Section 3 comprises exponential adaptive strategy-based SMOA (EASMO). Various numerical standard problems are used to investigate the performance of EASMO, narrated in Sect. 4. Conclusively, Sect. 5 winds up the work and assimilates the summary.

2 Spider Monkey Optimization Algorithm (SMOA)

This stochastic optimization method, proposed by Bansal et al. [1], is inspired by the social nature of spider monkeys (SMs). These SMs come in the class of fission–fusion social structure (FFSS)-based animals. In the food searching process, the FFSS-based animals separated themselves into subgroups. A female more often leads the group and if that female leader will not get plenteous food for that group, the group will be divided into small subgroups. Depending upon the accessibility of food sources, the members of the subgroups impart inside and outside [1].

At first, the population of N spider monkeys is uniformly distributed in the given search space as expressed into Eq. 1:

$$SM_{ij} = SM_{minj} + U(0,1) \times (SM_{maxj} - SM_{minj}) \tag{1}$$

here each monkey SM_i $(i = 1, 2, \ldots, N)$ is a D-dimensional vector and SM_{ij} specify the jth dimension of the ith spider monkey, SM_{minj} and SM_{maxj} are the lower limit and upper limit of the ith spider monkey in jth dimension and $U(0, 1)$ is the evenly assigned arbitrary values specified in the interval [0, 1].

The SMOA comprises six stages, namely Local Leader Stage, Global Leader Stage, Local Leader Learning Stage, Global Leader Learning Stage, Local Leader Decision Stage, and Global Leader Decision Stage. Firstly, like the other SI-based algorithms, the swarm of the SMOA is initialized in the given search space. After that, SMOA stages evolve the swarm to reach the global optima.

2.1 Local Leader Stage (LLS)

During this stage, the position of every SM is updated relied upon the local leader's position and the position of randomly selected neighbor within the group. Then, the fitness of the new solution is calculated, and by applying the greedy selection, the best fitted SM is adopted. The position update equation for ithSM (kth local group member) is expressed by Eq. 2:

$$SMnew_{ij} = SM_{ij} + U(0,1) \times (LL_{kj} - SM_{ij}) + U(-1,1) \times (SM_{rj} - SM_{ij}) \tag{2}$$

where SM_{ij} is the jth dimension of the ith spider monkey, SM_{rj} is the jth dimension of the arbitrary choosen rth spider monkey within the group k such that $r \neq i$. LL_{kj} is the jth dimension of the the kth local group leader.

2.2 Global Leader Stage (GLS)

After the execution of the LLS, GLS is executed. In the GLS, all the SMs are updated as per the knowledge received from the global leader and the members of the local group. The position update process takes place as per Eq. 3:

$$SMnew_{ij} = SM_{ij} + U(0,1) \times (GL_j - SM_{ij}) + U(-1,1) \times (SM_{rj} - SM_{ij}) \tag{3}$$

where GL_j depicts the position of the global leader in the jth direction. The term $j \in \{1, 2, ..., D\}$ is the arbitrarily choosen index. In this stage, more chances are given to the favorable solutions to make them better and for this purpose, a probability factor $prob_i$ is introduced. Here, the position of the SMs is updated on the basis of probability. The $prob_i$ is calculated by using the following equation:

$$prob_i = 0.9 \times \frac{fitness_i}{max\_fitness} + 0.1 \tag{4}$$

here, $max\_fitness$ is the maximum fitness, and $fitness_i$ is the fitness of ith SM of the group.

2.3 Global Leader Learning Stage (GLLS)

Here, the greedy selection method is employed on the population for updating the global leader's position. If the global leader's position is not updated up to a predetermined number of times, then the GlobalLimitCount is raised by 1.

2.4 Local Leader Learning Stage (LLLS)

The global leader's position is updated by applying the greedy selection method. Furthermore, the LocalLimitCount is raised by 1 when the local leader is not updating its location.

2.5 Local Leader Decision Stage (LLDS)

If a group's local leader is not updating its position according to a predestined threshold called LocalLeaderLimit, then by doing arbitrary initialization or by using the global and local leader's information, the position of the individuals from that group is updated.

2.6 Global Leader Decision Stage (GLDS)

If the position of the global leader is not updating up to a predetermined threshold extent known as GlobalLeaderLimit, then the global leader takes a decision and the population is split in the form of subgroups. The limit for the maximum number of small groups is defined by MG which is defined in Sect. 4.

3 Exponential Adaptive Strategy-Based SMOA

For the efficient performance of any population-based optimization algorithm, exploration and exploitation should be well balanced [3, 6, 9, 18]. In Local Leader Phase (LLP), the new solution is updated by learning from its previous experience as well as learning from the local leaders and the local group members [1]. Here, the chances for skipping the true solutions and stuck into the local optima are high. So balancing of step size should be managed in a way so that the chances of getting success in achieving the global optima will become maximum. Therefore to enhance the exploitation capability of SMOA, an exponential adaptive strategy [17] is applied in the LLS of SMOA. Equation 2 of LLS is modified as per Eq. 5.

$$SMnew_{ij} = SM_{ij} + \alpha(iteration) \times (LL_{kj} - SM_{ij}) + U(-1, 1) \times (SM_{rj} - SM_{ij}) \tag{5}$$

and the step size $\alpha(iteration)$ during each iteration is calculated by using Eq. 6.

$$\alpha(iteration) = \alpha_{max}\left(\frac{itr_{max} - iteration + 1}{itr_{max}}\right)^{\theta} \tag{6}$$

$$\theta = \frac{log(\frac{\alpha_{min}}{\alpha_{max}})}{log(\frac{1}{itr_{max}})} \tag{7}$$

here, α_{max} and α_{min} are the maximum and minimum step size reduction coefficient, itr_{max} is the maximum generation number, and iteration is the current generation number of the algorithm. θ is a coefficient, and its value is calculated by using Eq. 7.

In the first few iterations, α enhance the exploration capabilities by having larger values. As iteration increases, α decreases. Hence, it will decrease the step size exponentially for further iterations and improves the exploitation capability as well as the convergence speed. In the developed strategy, only the LLS is modified, while other stages are kept the same as mentioned in the basic SMOA. The sensitivity analysis of α with respect to iteration is depicted in Fig. 1. The graph depicts that the value of α decreases with the increase in the number of iterations.

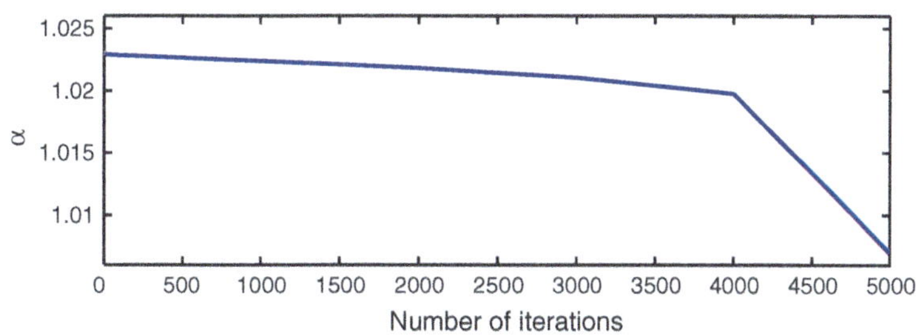

Fig. 1 Graph between α and the number of iterations

On the basis of the above clarification, modified LLS is depicted in Algorithm 1.

```
for each member SM_i ∈ kth group do
  for each j ∈ {1, ...., D} do
    if U(0, 1) ≥ pr then
      Calculate the value of θ as per equation 7
      Calculate the value of α(iteration) as per equation 6
      SMnew_ij = SM_ij + α(iteration) × (LL_kj − SM_ij) +
      U(−1, 1) × (SM_rj − SM_ij)
    else
      SMnew_ij = SM_ij
    end if
  end for
end for
```

Algorithm 1: Position Update Process in LLS

4 Discussion and Analysis

To investigate the pursuance of the propound EASMO, 14 distinctive global optimization problems (OP_1 to OP_{14}) are chosen as depicted in Table 1.

To approve the compatibility of the EASMO, a comparative analysis is carried out among EASMO, ABC [7], DE [18], GABC [24], PSO [19], SMOA [1], and three of its variant, namely LFSMOA [14], MLSMOA [15], and PLSMOA [13]. Experimental setup for the comparison is as follows:

- The numbers of simulations = 100,
- Population Size $N = 50$,
- $GlobalLeaderLimit = N$,
- $LocalLeaderLimit = D \times N$,
- $pr = 0.1 + (\frac{0.4}{Max\_iterations})$,
- $\alpha_{max} = 1.020$,
- $\alpha_{min} = 1.007$,
- $MG = N/10$,
- The parameter settings for SMOA [1], ABC [7], DE [18], GABC [24], PSO [19], LFSMOA [14], MLSMOA [15] and PLSMOA [13] are taken from the original research.
- To study the impact of the parameters α_{min} and α_{max} on the performance of EASMO, its susceptibility with respect to various values of α_{max} and α_{min} in the range of [0, 2], is examined in Figs. 2 and 3. Figures 2 and 3 show that the algorithm is exceptionally delicate toward α_{max} and α_{min} and their value 1.020 for α_{max} and 1.007 for α_{min} give comparatively better results.Therefore,α_{max} = 1.02 and α_{min} = 1.007 are choosed in this paper for the experiments.

Table 1 OP: Optimization problem, D: Dimension, AE: Acceptable error

Objective functions	Search range	Optimum value	D	AE
$OP_1(x)=\sum_{i=1}^{D}(100(x_{i+1}-x_i^2)^2+(x_{i-1})^2)$	[−30, 30]	$OP(\mathbf{1})=0$	30	1.0E−02
$OP_2(x)=\sum_{i=1}^{D}\lvert x_i\sin x_i+0.1x_i\rvert$	[−10, 10]	$OP(0)=0$	30	1.0E−05
$OP_3(x)=\sum_{i=1}^{D}{x_i}^2+(\sum_{i=1}^{D}\frac{ix_i}{2})^2+(\sum_{i=1}^{D}\frac{ix_1}{2})^4$	[−5.12, 5.12]	$OP(0)=0$	30	1.0E−02
$OP_4(x)=-\sum_{i=1}^{D-1}\left(\exp\left(\frac{-(x_i^2+x_{i+1}^2+0.5x_ix_{i+1})}{8}\right)\times I\right)$ where, $I=cos\left(4\sqrt{x_i^2+x_{i+1}^2+0.5x_ix_{i+1}}\right)$	[−5, 5]	$OP(0)=-D+1$	10	1.0E−05
$OP_5(x)=[1.5-x_1(1-x_2)]^2+[2.25-x_1(1-x_2^2)]^2+[2.625-x_1(1-x_2^3)]^2$	[−4.5, 4.5]	$OP(3,0.5)=0$	2	1.0E−05
$OP_6(x)=100[x_2-x_1^2]^2+(1-x_1)^2+90(x_4-x_3^2)^2+(1-x_3)^2+10.1[(x_2-1)^2+(x_4-1)^2]+19.8(x_2-1)(x_4-1)$	[−10, 10]	$OP(1)=0$	4	1.0E−05
$OP_7(x)=\sum_{i=1}^{D}z_i^2+f_{bias},\ z=x-o\ ,x=[x_1,x_2,....x_D],\ o=[o_1,o_2,...o_D]$	[−100, 100]	$OP(o)=OP_{bias}=-450$	10	1.0E−05
$OP_8(x)=-\cos x_1\cos x_2e^{(-(x_1-\pi)^2-(x_2-\pi)^2)}$	[−100, 100]	$OP(\pi,\pi)=-1$	2	1.0E−13
$OP_9(x)=10^5x_1^2+x_2^2-(x_1^2+x_2^2)^2+10^{-5}(x_1^2+x_2^2)^4$	[−20, 20]	$OP(0,15)=F(0,-15)=-24777$	2	5.0E−01
$OP_{10}(x)=\sin(x_1+x_2)+(x_1-x_2)^2-\frac{3}{2}x_1+\frac{5}{2}x_2+1$	$-1.5\le x_1\le 4, -3\le x_2\le 3$	$OP(-0.547,-1.547)=-1.9133$	2	1.0E−04
$OP_{11}(x)=-\sum_{i=1}^{5}i\cos((i+1)x_1+1)\sum_{i=1}^{5}i\cos((i+1)x_2+1)$	[−10, 10]	$OP(7.0835,4.8580)=-186.7309$	2	1.0E−05
$OP_{12}(x)=-[A\prod_{i=1}^{D}\sin(x_i-z)+\prod_{i=1}^{D}sin(B(x_i-z))]$, $A=2.5, B=5, z=30$	[0, 180]	$OP(90+z)=-(A+1)$	10	1.0E−02
$OP_{13}(x)=4\varepsilon[(\frac{\sigma}{\gamma})^{12}-(\frac{\sigma}{\gamma})^6]=\varepsilon[(\frac{\gamma m}{\gamma})^{12}-2(\frac{\gamma m}{\gamma})^6]$	[−2, 2]	−9.103852	15	1.0E−03
$OP_{14}(x)=-\sum_{i=1}^{D}x_i sin(\sqrt{\lvert x_i\rvert})$	[−100, 100]	$OP(0)=0$	30	1.0E−03

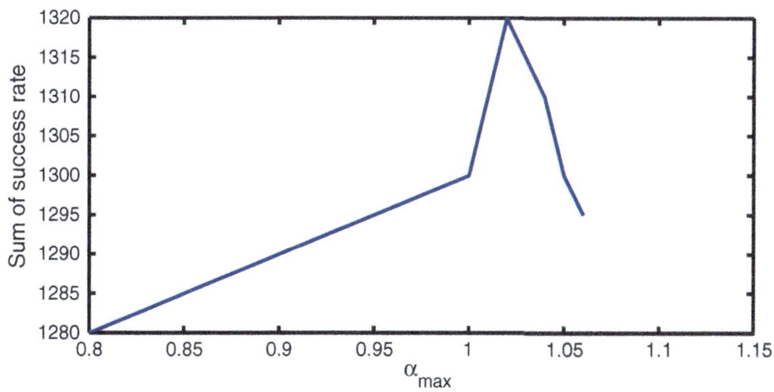

Fig. 2 Effect of parameter α_{max} on success rate

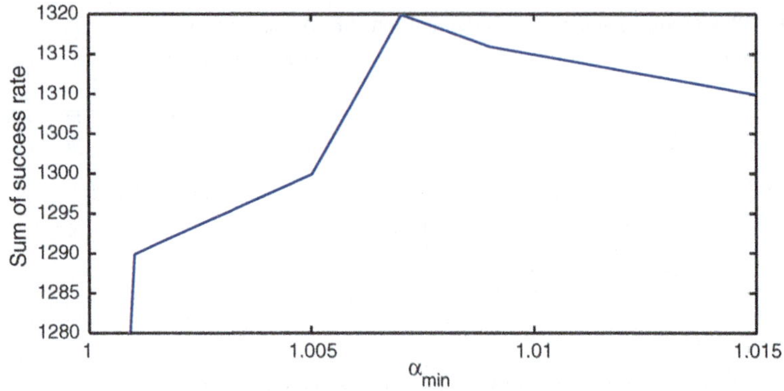

Fig. 3 Effect of parameter α_{min} on success rate

The experimental outcomes for the considered algorithms are exhibited in Table 2. Table 2 represents the results in the form of standard deviation (SD), average count of function evaluations ($AFEs$), mean error (ME) along with the success rate (SR). Results in Table 2 reveal that the EASMO outplays in the form of efficiency, accuracy as well as reliability in comparison with the considered algorithms.

Table 2 Collation of the outcome of optimization problem, OP: Optimization problem

OP	Algorithm	SD	ME	AFEs	SR
OP_1	EASMO	1.28E+05	2.66E+04	327757.89	51
	SMOA	2.77E+01	2.04E+01	507418.68	5
	ABC	2.59E+00	9.58E−01	240192.00	10
	GABC	9.73E+00	4.48E+00	243403.08	6
	DE	6.24E+01	3.62E+01	250050.00	0
	PSO	4.75E+01	4.07E+01	250050.00	0
	LFSMOA	4.36E+01	4.81E+01	200069.05	0
	MLSMOA	2.18E+03	1.41E+03	200495.00	0
	PLSMOA	3.70E+01	3.43E+01	200056.07	0
OP_2	EASMO	2.40E−06	8.12E−06	69252.84	100
	SMOA	6.19E−07	9.44E−06	83959.22	100
	ABC	1.80E−06	8.30E−06	76266.50	100
	GABC	2.18E−06	7.97E−06	57913.04	100
	DE	9.26E−07	9.30E−06	63978.50	100
	PSO	4.00E−07	9.58E−06	94751.50	100
	LFSMOA	3.38E−04	1.23E−04	161894.07	60
	MLSMOA	3.83E−01	4.55E−01	200495.00	0
	PLSMOA	5.22E−04	1.21E−04	139888.24	68

(continued)

Table 2 (continued)

OP	Algorithm	SD	ME	AFEs	SR
OP_3	EASMO	8.89E−04	9.29E−03	91321.87	100
	SMOA	8.03E−04	9.33E−03	132125.82	100
	ABC	1.44E+01	9.10E+01	250000.00	0
	GABC	1.76E+01	9.16E+01	250000.00	0
	DE	1.19E−03	9.10E−03	71045.50	100
	PSO	5.37E−04	9.58E−03	207048.00	100
	LFSMOA	7.84E−04	9.32E−03	161360.01	100
	MLSMOA	1.21E−02	3.48E−02	200495.00	0
	PLSMOA	5.65E−04	9.61E−03	115965.19	100
OP_4	EASMO	3.67E−06	5.58E−06	45487.03	100
	SMOA	1.70E−06	8.06E−06	79609.01	100
	ABC	1.60E−04	2.43E−05	91044.53	97
	GABC	2.42E−06	7.10E−06	44735.63	100
	DE	6.48E−01	9.74E−01	219967.00	16
	PSO	6.40E−01	1.32E+00	243330.00	5
	LFSMOA	1.08E−01	1.33E−02	85488.54	95
	MLSMOA	8.60E−01	6.17E−01	199695.00	2
	PLSMOA	9.59E−02	1.34E−02	69924.43	97
OP_5	EASMO	2.99E−06	4.55E−06	1054.35	100
	SMOA	2.97E−06	4.80E−06	1478.07	100
	ABC	1.62E−06	8.73E−06	16843.86	100
	GABC	2.88E−06	5.51E−06	9062.85	100
	DE	6.24E−07	4.28E−06	1871.00	100
	PSO	3.11E−06	4.88E−06	2807.50	100
	LFSMOA	2.58E−06	7.55E−06	1345.29	100
	MLSMOA	2.42E−06	1.89E−06	49019.78	100
	PLSMOA	2.98E−06	4.95E−06	1418.78	100
OP_6	EASMO	1.35E−04	9.04E−04	21736.95	100
	SMOA	2.39E−04	7.57E−04	53552.63	100
	ABC	9.34E−02	1.45E−01	250151.98	0
	GABC	1.08E−02	1.25E−02	250045.38	1
	DE	4.77E−01	1.09E−01	40128.00	86
	PSO	1.58E−04	8.39E−04	49970.50	100
	LFSMOA	3.66E−05	9.80E−04	17656.19	100
	MLSMOA	6.84E−02	1.00E−01	196488.07	3
	PLSMOA	1.70E−04	8.68E−04	18280.78	100

(continued)

Table 2 (continued)

OP	Algorithm	SD	ME	AFEs	SR
OP_7	EASMO	1.60E−06	7.61E−06	4740.12	100
	SMOA	1.92E−06	7.16E−06	5936.04	100
	ABC	2.61E−06	6.66E−06	9004.50	100
	GABC	2.24E−06	6.93E−06	5587.00	100
	DE	1.31E−06	7.68E−06	10895.00	100
	PSO	1.50E−06	8.37E−06	15822.50	100
	LFSMOA	1.70E−06	7.79E−06	7461.81	100
	MLSMOA	3.02E−04	4.34E−04	200495.00	0
	PLSMOA	1.75E−06	7.61E−06	6141.89	100
OP_8	EASMO	2.77E−14	4.48E−14	6835.17	100
	SMOA	2.92E−14	4.63E−14	12299.03	100
	ABC	3.49E−05	1.11E−05	243317.32	14
	GABC	5.37E−09	5.40E−10	45017.81	99
	DE	1.29E−14	4.95E−14	5203.00	100
	PSO	2.70E−14	5.35E−14	9867.00	100
	LFSMOA	2.93E−14	5.36E−14	12942.91	100
	MLSMOA	3.44E−06	2.16E−06	200495.00	0
	PLSMOA	2.85E−14	5.22E−14	12459.74	100
OP_9	EASMO	5.55E−03	4.89E−01	955.35	100
	SMOA	4.98E−03	4.90E−01	1192.95	100
	ABC	5.54E−03	4.89E−01	1432.56	100
	GABC	5.22E−03	4.89E−01	789.01	100
	DE	2.09E−03	4.90E−01	2738.50	100
	PSO	5.54E−03	4.92E−01	4723.50	100
	LFSMOA	5.66E−03	4.91E−01	900.26	100
	MLSMOA	3.93E−03	4.85E−01	44207.90	100
	PLSMOA	5.48E−03	4.90E−01	1214.17	100
OP_{10}	EASMO	6.49E−06	8.62E−05	548.46	100
	SMOA	6.53E−06	8.72E−05	766.26	100
	ABC	6.46E−06	8.87E−05	1193.07	100
	GABC	6.70E−06	8.88E−05	624.50	100
	DE	2.18E−06	8.78E−05	1419.50	100
	PSO	6.39E−06	8.87E−05	1507.50	100
	LFSMOA	6.53E−06	8.90E−05	695.81	100
	MLSMOA	2.73E−08	7.71E−05	40198.00	100
	PLSMOA	7.48E−06	8.81E−05	714.54	100

(continued)

Table 2 (continued)

OP	Algorithm	SD	ME	AFEs	SR
OP_{11}	EASMO	5.39E−06	4.68E−06	2362.14	100
	SMOA	5.68E−06	4.95E−06	4313.43	100
	ABC	5.55E−06	4.86E−06	4633.30	100
	GABC	5.77E−06	5.14E−06	2360.72	100
	DE	2.69E−06	4.40E−06	9562.50	100
	PSO	2.11E−03	4.29E−04	103221.50	71
	LFSMOA	5.94E−06	5.26E−06	2300.62	100
	MLSMOA	2.53E−05	1.08E−05	113565.31	86
	PLSMOA	5.93E−06	5.34E−06	3865.29	100
OP_{12}	EASMO	2.41E−03	7.99E−03	143305.00	95
	SMOA	4.22E−03	9.34E−03	236620.15	81
	ABC	1.96E−03	8.06E−03	54974.74	92
	GABC	2.35E−03	7.90E−03	49857.62	99
	DE	2.43E−01	3.13E+00	248696.00	1
	PSO	2.83E−01	3.74E−01	212422.00	25
	LFSMOA	3.61E−03	8.77E−03	83137.93	90
	MLSMOA	2.36E−03	6.03E−03	103953.43	98
	PLSMOA	1.27E−03	8.64E−03	27219.35	100
OP_{13}	EASMO	1.69E−04	8.20E−04	38951.83	100
	SMOA	5.98E−04	1.09E−03	274879.26	74
	ABC	1.25E−04	8.75E−04	70489.94	100
	GABC	4.11E−04	1.01E−03	116653.35	78
	DE	1.99E−04	7.13E−04	68619.50	100
	PSO	2.85E−01	1.69E−01	172325.00	54
	LFSMOA	4.37E−04	9.42E−04	57499.78	98
	MLSMO	1.87E−01	2.63E−02	54630.67	97
	PLSMOA	8.81E−05	9.17E−04	32449.10	100
OP_{14}	EASMO	1.45E−01	3.55E−02	215547.40	77
	SMOA	3.54E−01	8.22E−02	249974.06	75
	ABC	7.78E−02	2.18E−01	250000.21	0
	GABC	2.33E−03	9.89E−03	250000.00	0
	DE	4.71E+00	1.16E+01	250050.00	0
	PSO	3.11E−02	2.67E−02	250050.00	0
	LFSMOA	1.09E+00	6.74E−01	188933.23	22
	MLSMOA	0.00E+00	9.10E+00	200495.00	0
	PLSMOA	1.75E+00	1.15E+00	180908.97	26

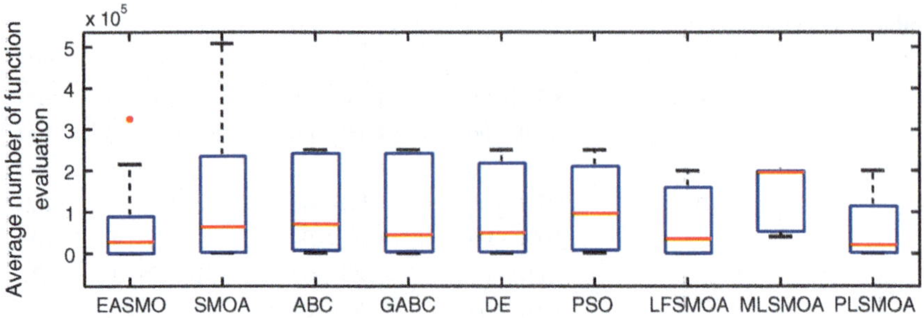

Fig. 4 Boxplot graph for average count of *function* evaluation

To measure the efficiency of the EASMO, boxplot (bp) analysis [16] is carried out on the basis of AFEs as shown in Fig. 4. The results reveal that the interquartile range and medians of EASMO are comparably low to the considered algorithms.

Further, acceleration rate (AR) is calculated for the considered algorithms in comparsion with EASMO using Eq. 8. The calculated AR is shown in Table 3. It is clear from Table 3 that the EASMO is comparatively fast than the other considered algorithms. AR is calculated as:

$$AR = \frac{AFE_{ALGO}}{AFE_{EASMO}} \tag{8}$$

Table 3 Acceleration rate of EASMO compared to the SMOA, ABC, GABC, DE, PSO, LFSMOA, MLSMOA, and PLSMOA

Test problems	SMOA	ABC	GABC	DE	PSO	LFSMOA	MLSMOA	PLSMOA
fn_1	1.54	0.73	0.74	0.76	0.76	0.61	0.61	0.61
fn_2	1.21	1.10	0.83	0.92	1.36	2.33	2.89	2.01
fc_3	1.44	2.73	2.73	0.77	2.26	1.76	2.19	1.26
fn_4	1.75	2.00	0.98	4.83	5.34	1.87	4.39	1.53
fn_5	1.40	15.97	8.59	1.77	2.66	1.27	46.49	1.34
fn_6	2.46	11.50	11.50	1.84	2.29	0.81	9.03	0.84
fn_7	1.25	1.89	1.17	2.29	3.33	1.57	42.29	1.29
fn_8	1.79	35.59	6.58	0.76	1.44	1.89	29.33	1.82
fn_9	1.24	1.49	0.82	2.86	4.94	0.94	46.27	1.27
fn_{10}	1.39	2.17	1.13	2.58	2.74	1.26	73.29	1.30
fn_{11}	1.8	1.96	0.99	4.04	43.69	0.97	48.07	1.63
fn_{12}	1.65	0.38	0.34	1.73	1.48	0.58	0.72	0.18
fn_{13}	7.05	1.80	2.99	1.76	4.42	1.47	1.40	0.83
fn_{14}	1.15	1.15	1.15	1.16	1.16	0.87	0.93	0.83

Table 4 Collation based on AFE and the MVUR sum test at ('+' indicates EASMO is better, '–' shows EASMO is worse, and '=' shows that there is no noticeable distinction)

Function no	EASMO versus SMOA	EASMO versus ABC	EASMO versus GABC	EASMO versus DE	EASMO versus PSO	EASMO versus LFSMOA	EASMO versus MLSMOA	EASMO versus PLSMOA
OP_1	+	–	–	–	–	–	–	–
OP_2	+	+	–	–	+	+	+	+
OP_3	+	+	+	–	+	+	+	+
OP_4	+	+	–	+	+	+	+	+
OP_5	+	+	+	+	+	+	+	+
OP_6	+	+	+	+	+	–	+	–
OP_7	+	+	+	+	+	+	+	+
OP_8	+	–	+	–	+	+	+	+
OP_9	+	+	–	+	+	–	+	+
OP_{10}	+	+	+	+	+	+	+	+
OP_{11}	+	+	–	+	+	–	+	+
OP_{12}	+	–	–	+	+	–	–	–
OP_{13}	+	+	+	+	+	+	+	–
OP_{14}	+	+	+	+	+	–	–	–
Total number of '+' sign	14	11	8	10	13	8	11	9

Mann–Whitney U rank (MWUR) sum test based on mean function evaluations at $\alpha = 0.005$ significance level is also done as shown in Table 4. Table 4 depicts that for most of the standard problems, EASMO perform outstanding. While analyzing the results of Table 4, it can be stated that the EASMO performs well on all the 14 optimization problems as compared to SMOA, on 11 problems ($OP_2 - OP_7$, $OP_9 - OP_{11}$, $OP_{13} - OP_{14}$) as compared to ABC, on 8 problems (OP_3, $OP_5 - OP_8$, OP_{10}, $OP_{13} - OP_{14}$) as compared to GABC, on 10 problems ($OP_4 - OP_7$, $OP_9 - OP_{14}$) as compared to DE, on 13 problems ($OP_2 - OP_{14}$) as compared to PSO, on 8 problems ($OP_2 - OP_5$, $OP_7 - OP_8$, OP_{10}, OP_{13}) as compared to LFSMO, on 11 problems ($OP_2 - OP_{11}$, OP_{13}) as compared to MLSMO, and on 9 problems ($OP_2 - OP_5$, $OP_7 - OP_{11}$) as compared to PLSMO.

5 Conclusion

In this article, a new variant of SMOA is developed, namely exponential adaptive spider monkey optimization (EASMO) algorithm. In the Local Leader Stage, an exponential adaptive strategy is applied to balance the step size with respect to the iterations, which enhance the exploitation ability of the SMOA. Further, the EASMO has analyzed over 14 standard optimization problems and the obtained results are compared with eight significant algorithms. Through Mann–Whitney U rank sum test, boxplot analysis, and acceleration rate, it is proved that the propound algorithm is an efficient SI-based algorithm. In the future, the propound algorithm may be applied to solve complex engineering optimization problems.

References

1. J.C. Bansal, H. Sharma, S.S. Jadon, C. Maurice, Spider monkey optimization algorithm for numerical optimization. Memet. Comput. **6**(1), 31–47 (2014)
2. M. Dorgio, T. Stutzle, *Ant Colony Optimization*, A Bradferd Book (MCT Press, England, 2004)
3. R. Eberhart, J. Kennedy, A new optimizer using particle swarm theory, in *1995 Proceedings of the Sixth International Symposium on Micro Machine and Human Science, MHS'95* (IEEE, 1995), pp. 39–43
4. D.B. Fogel. *Evolutionary Computation: The Fossil Record* (Wiley-IEEE Press, 1998)
5. R.A. Formato, Central force optimization: a new nature inspired computational framework for multidimensional search and optimization, in *Nature Inspired Cooperative Strategies for Optimization (NICSO 2007)* (Springer, 2008), pp. 221–238
6. D.E. Goldberg, J.H. Holland, Genetic algorithms and machine learning. Mach. Learn. **3**(2), 95–99 (1988)
7. D. Karaboga, B. Basturk, A powerful and efficient algorithm for numerical function optimization: artificial bee colony (ABC) algorithm. J. Global Optim. **39**(3), 459–471 (2007)
8. J. Kennedy, Particle swarm optimization, in *Encyclopedia of Machine Learning* (Springer, 2011), pp. 760–766
9. K.M. Passino, Biomimicry of bacterial foraging for distributed optimization and control. IEEE Control Syst. **22**(3), 52–67 (2002)
10. K.M. Passino, Bacterial foraging optimization. Int. J. Swarm Intell. Res. (IJSIR) **1**(1), 1–16 (2010)
11. K. Price, R.M. Storn, J.A. Lampinen, *Differential Evolution: A Practical Approach to Global Optimization* (Springer Science & Business Media, 2006)
12. V.S. Venkateswara Rao, R.S. Shekhawat, V.K. Srivastava, A reliable digital image watermarking scheme based on SVD and particle swarm optimization, in *2012 Students Conference on Engineering and Systems (SCES)* (IEEE, 2012), pp. 1–6
13. A. Sharma, H. Sharma, A. Bhargava, N. Sharma, Power law-based local search in spider monkey optimisation for lower order system modelling. Int. J. Syst. Sci. **48**(1), 150–160 (2017)
14. A. Sharma, H. Sharma, A. Bhargava, N. Sharma, J.C. Bansal, Optimal power flow analysis using lévy flight spider monkey optimisation algorithm. Int. J. Artif. Intell. Soft Comput. **5**(4), 320–352 (2016)
15. A. Sharma, H. Sharma, A. Bhargava, N. Sharma, J.C. Bansal, Optimal placement and sizing of capacitor using limaçon inspired spider monkey optimization algorithm. Memet. Comput. **9**(4), 311–331 (2017)
16. A. Sharma, H. Sharma, A. Bhargava, N. Sharma, J.C. Bansal, Black hole artificial bee colony algorithm, in *International Conference on Swarm, Evolutionary, and Memetic Computing* (Springer, 2015), pp. 214–221
17. A. Soltanian, F. Derakhshan, M. Soleimanpour-Moghadam, MWWO: modified water wave optimization, in *2018 3rd Conference on Swarm Intelligence and Evolutionary Computation (CSIEC)* (IEEE, 2018), pp. 1–5
18. R. Storn, K. Price, Differential evolution-a simple and efficient heuristic for global optimization over continuous spaces. J. Global Optim. **11**(4), 341–359 (1997)

19. Ioan Cristian Trelea, The particle swarm optimization algorithm: convergence analysis and parameter selection. Inf. Process. Lett. **85**(6), 317–325 (2003)
20. J. Vesterstrom, R. Thomsen, A comparative study of differential evolution, particle swarm optimization, and evolutionary algorithms on numerical benchmark problems. IEEE Congress Evol. Comput. **2**, 1980–1987 (2004)
21. G. Vinod, H.S. Kushwaha, A.K. Verma, A. Srividya, Optimisation of ISI interval using genetic algorithms for risk informed in-service inspection. Reliab. Eng. Syst. Saf. **86**(3), 307–316 (2004)
22. X.-S. Yang, A.H. Gandomi, S. Talatahari, A.H. Alavi. *Metaheuristics in Water, Geotechnical and Transport Engineering* (Newnes, 2012)
23. X.-S. Yang, M. Karamanoglu, Swarm intelligence and bio-inspired computation: an overview, in *Swarm Intelligence and Bio-Inspired Computation* (Elsevier, 2013), pp. 3–23
24. G. Zhu, S. Kwong, Gbest-guided artificial bee colony algorithm for numerical function optimization. Appl. Math. Comput. **217**(7), 3166–3173 (2010)

Development of Fuzzy Knowledge-Based System for Water Quality Assessment in River Ganga

Praveen Kumar Shukla

Abstract Rivers are playing an important role in human life and wildlife but due to pollution the quality of river water is extremely deteriorated. The assessment of water quality is a very indeterminate task and associates a lot of uncertainty and subjectivity in the decision making. To cope with this situation, computational intelligence techniques are found competent to develop models for water quality assessment. One of the computational intelligence techniques, fuzzy logic is used to implement such models. In this manuscript, a fuzzy knowledge-based system is developed to classify the water quality of river Ganga in three groups. The open-access software 'Guaje' is used to implement the proposed model. The analysis of the results is presented in terms of interpretability and accuracy which are found satisfactory.

Keywords Fuzzy rule-based system · Fuzzy knowledge-based system · Fuzzy logic

1 Introduction

Rivers are the crucial natural resource of any nation playing a big role in the human life and wildlife as well as in the economy of a country by helping in the agriculture and industry. Due to pollution, the water quality degraded exponentially resulting in many problems in the human life and wildlife. Considering the above reason, it is really very important to have some mechanism for assessing river water quality. The assessment of the water quality of any river is extremely indeterminate due to continuous change in the values of decision parameters [1]. Several parameters are used for assessing the quality of the water. Few of the parameters are color, odor, total dissolved solids (TDS), total suspended solids (TSS), pH, chemical oxygen demand (COD), biochemical oxygen demand (BOD), turbidity, electrical conductivity (EC),

P. K. Shukla (✉)
Department of Information Technology, Babu Banarsi Das Northern India Institute of Technology, Lucknow, India
e-mail: drpraveenkumarshukla@gmail.com

© Springer Nature Singapore Pte Ltd. 2020

A. K. Nagar et al. (eds.), *Soft Computing for Problem Solving 2019*,
Advances in Intelligent Systems and Computing 1139,
https://doi.org/10.1007/978-981-15-3287-0_2

dissolved oxygen (DO) and salinity. Sometimes, in the water toxic elements, pesticides and anthropogenic sources are found in the water due to the reasons of the pollution. In the laboratory, the other parameters which are measured to test the water quality are nitrate (NO^{2-} and NO^{3-}), ammonia, sulfates, chloride and calcium.

The task of water quality evaluation and classification requires a competent model for precise and real-time decision making [2]. To deal with the uncertainty and subjectivity, in this case, it is recommended to use fuzzy techniques which lead to the development of fuzzy rule-based systems [25] for water quality assessment [3, 4].

The development of FRBS has two important parameters; accuracy and interpretability [21, 22]. Accuracy measures the closeness between a modeled system and real system. Normally, it is measured in terms of mean square error (MSE) or root mean square error (RMSE). On the other hand, the interpretability is the measure of understating a system by inspecting its knowledge base [23, 24]. The interpretable systems are the simple systems which are not complex. The interpretability and accuracy have a trade-off relation [26], i.e., more interpretable systems are less accurate and more accurate systems are less interpretable due to its increased complexity [27, 28]. In the development of this model, both the parameters are recorded and analyzed.

The paper consists of five sections. Here, Sect. 2 introduces the related work and the proposed work is introduced in Sect. 3. Section 4 is the implementation of the proposed model. The conclusion and future scope is given in Sect. 5.

2 Related Work

Several proposals and implementations are found in the literature to assess the water quality using fuzzy logic which is briefly reviewed below in this section.

An index of water quality assessment using fuzzy logic is developed in [5, 8]. The parameters related to physicochemical, toxicological and microbial areas are included in the index development. The subjectivity of the water quality index is dealt with using fuzzy logic.

A fuzzy inference system-based water quality index is proposed in [6]. A multi-attribute decision making is also used in this quality index. Another water quality index is developed for underground water in [7].

The water quality management policies are implemented using fuzzy logic-based approach in [9] which is based on multiple attribute decision making.

The other related works are outlined in Table 1.

3 Proposed Model

The fuzzy rule-based system is proposed to classify the river water into three classes; Class A, Class B, Class C. This classification is based on the four input parameters;

Table 1 Fuzzy logic used in water quality index assessment

S. no.	Work done	References
1	Evaluation of groundwater quality to find the suitability of drinking water	[10, 19]
2	A fuzzy model of water quality assessment for the rivers of India and Malaysia	[11, 13, 21]
3	Assessment of dairy cattle drinking water quality index (DCWQI) which is fuzzy inference systems based	[12]
4	Fuzzy river water quality in Morocco	[14]
5	Development of water quality in shrimp ecosystem using fuzzy inference system considering chemical, physical and biological parameters	[15]
6	Water quality evaluation using hybrid fuzzy and optimal modeling	[16]
7	Fuzzy water quality index for Sorocaba river	[17]
8	Water quality index based on cascaded fuzzy systems	[18]
9	Comparison of river water quality assessment approach based on binary and fuzzy logic	[20]

Fig. 1 Proposed fuzzy rule-based system

dissolved oxygen (DO), biochemical oxygen demand (BOD), pH, total coliform. The block diagram of the proposed model is given in Fig. 1.

The descriptions of the inputs are as follows;

1. **DO**: This is the amount of gaseous oxygen which is dissolved in the water. This amount is affected by the temperature of the water and the volume of the moving water. The dissolved oxygen is very crucial for aquatic creatures to respire.
2. **BOD**: This is the amount of dissolved oxygen required by aerobic biological organisms for breaking down the organic material available in a prescribed water sample considering the specific temperature over prescribed time period.
3. **pH**: This is a scale to know the nature of a water solution; acidic or basic. The pH scale ranges from 1 to 14. The water with pH value from 6 to 8.5 is safe for drinking.
4. **Total Coliform**: These are the group of similar types of bacteria which are not harmful. So, the count of such bacteria is used to check the presence of other harmful bacteria in the water.

At the output, three classes are as follows,

1. **Class A**: Fit for drinking after disinfection.
2. **Class B**: Fit for drinking with conventional treatment after disinfection.

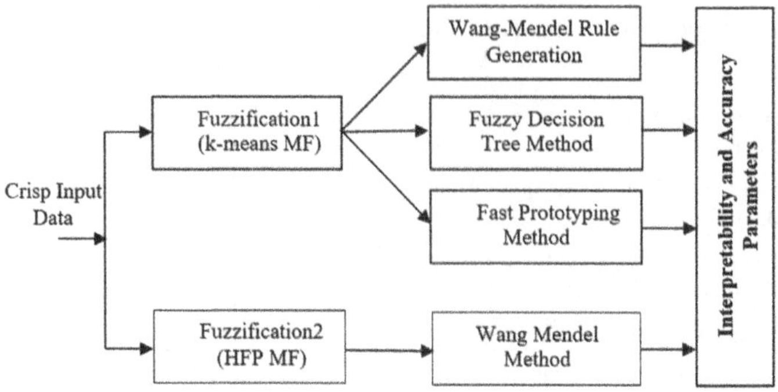

Fig. 2 Implementation mechanism

Table 2 Fuzzy partition and rule generation methods for experiments

Experiments	Fuzzy partition method	Rule generation method
Experiment-1	k-means MF	Wang–Mendel method
Experiment-2	k-means MF	Fuzzy decision tree method
Experiment-3	k-means MF	Fast prototyping algorithm
Experiment-4	HFP MF	Wang–Mendel method

3. **Class C**: Fit for drinking only after organized conventional/advance treatment including disinfection.

The strategy of the implementation of the proposed FRBS is discussed in Fig. 2.

For the proposed model, four experiments (refer Table 2) are designed which are discriminated with fuzzy partition technique and rule generation method (Fig. 2).

4 Implementation and Result Analysis

The water quality classification is implemented using the data set obtained from the Web site of Nmami Gange Yojna of the Govt. of India [29]. The sample has been taken from the 32 stations of Uttar Pradesh and Bihar. The stations of Uttar Pradesh are included from Rishkesh, Harki Pauri, Haridwar, Gangotri, Garhmukteshwar, Kannuaj, Bithoor, Raebareli, Prayag, Varanasi and Mirzapur. Chapra, Patna and Buxar are the important station areas from Bihar.

Guaje open-access software [30] is used to implement the proposed model. The membership functions for Experiment 1, 2 and 3 are given in Figs. 3, 4, 5 and 6, respectively.

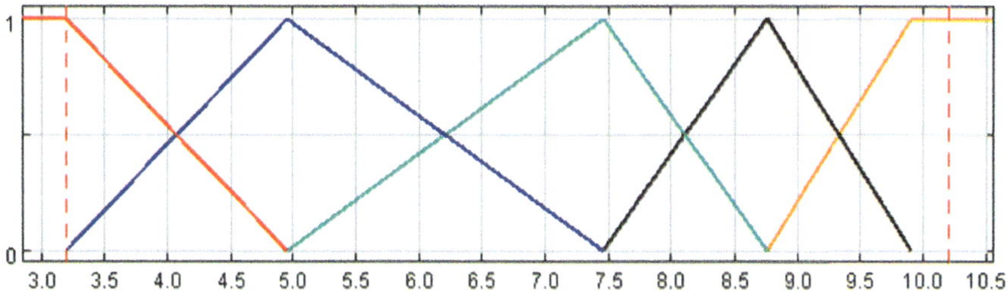

Fig. 3 Membership functions of dissolved oxygen

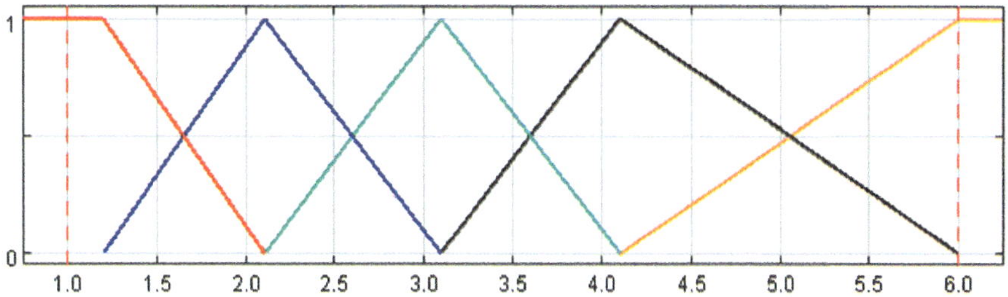

Fig. 4 Membership functions of BOD

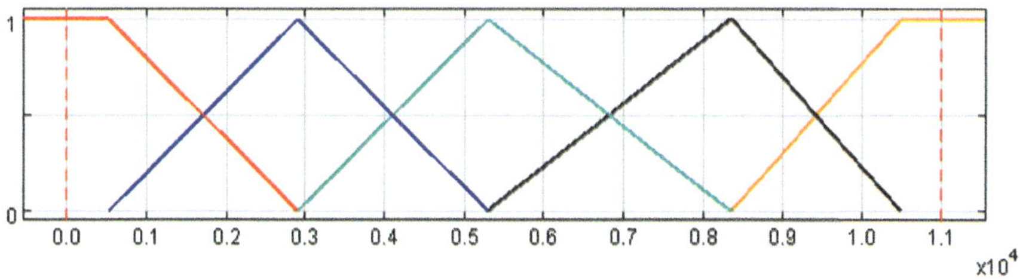

Fig. 5 Membership function of total coliform

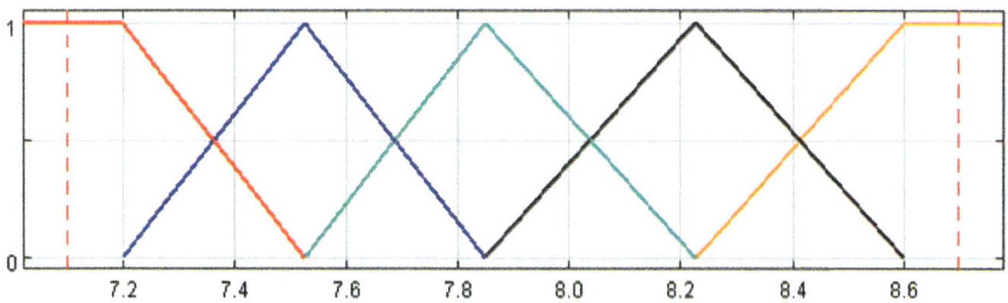

Fig. 6 Membership function of pH

Table 3 Accuracy and interpretability parameters in Experiment-1

Accuracy parameters		Interpretability parameters	
Accuracy (%)	77.4%	Nauck's index	0.023
MSE	0.095	NOR	16
		TRL	64
		ARL	4
		Average inferential fired rule	5.857
		Accumulated rule complexity	16.088

Experiment-1
Method of Rule Generation: Wang and Mendel (Table 3).

Experiment-2
Method of Rule Generation: Fuzzy Decision Tree.

The settings of the experiment are as follows,
Maximum tree depth = 5
Maximum significant level = 0.2
Leaf minimum cardinality = 10
Tolerance threshold = 0.1
Minimum entropy deviance gain = 0.001
Coverage threshold = 0.9
Pruning = Yes (Table 4).

Experiment-3
Method of Rule Generation: Fast Prototyping Algorithm.
The settings of the experiment are as follows,

Strategy-Decreasing
Minimum cardinality = 3
Minimum degree = 0.3 (Table 5).

Table 4 Accuracy and interpretability parameters in Experiment-2

Accuracy parameters		Interpretability parameters	
Accuracy (%)	67.7%	Nauck's index	0.188
MSE	0.046	NOR	5
		TRL	8
		ARL	1.6
		Average inferential fired rule	5.011
		Accumulated rule complexity	2.516

Table 5 Accuracy and interpretability parameters in Experiment-3

Accuracy parameters		Interpretability parameters	
Accuracy (%)	61.3%	Nauck's Index	0.022
MSE	0.131	NOR	17
		TRL	68
		ARL	4
		Average inferential fired rule	17.093
		Accumulated rule complexity	9.828

Experiment-4

The membership functions are changed into HFP with number of partitions 5 (Figs. 7, 8, 9 and 10). We found the following observations in this experiment.

Accuracy Parameters

Accuracy = 93.5
MSE = 0.063

Interpretability Parameters

NOR = 27
TRL = 108.

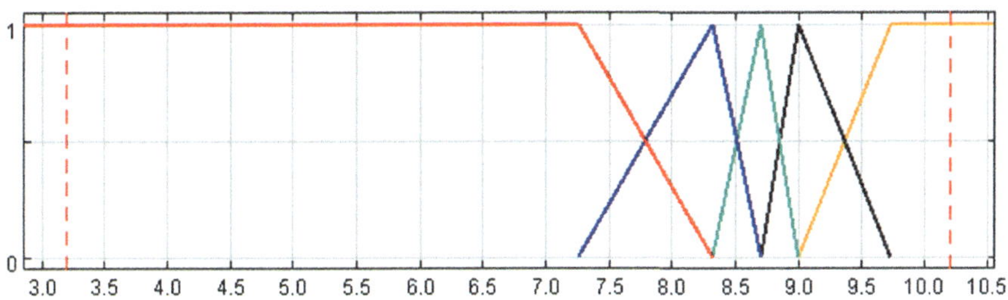

Fig. 7 Membership function of dissolved oxygen

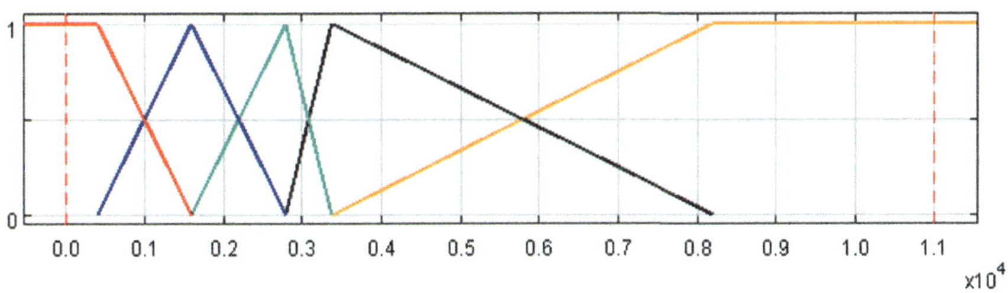

Fig. 8 Membership function of BOD

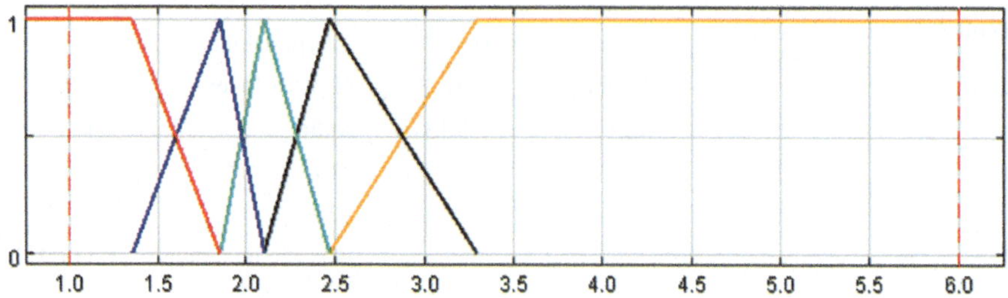

Fig. 9 Membership function of total coliform

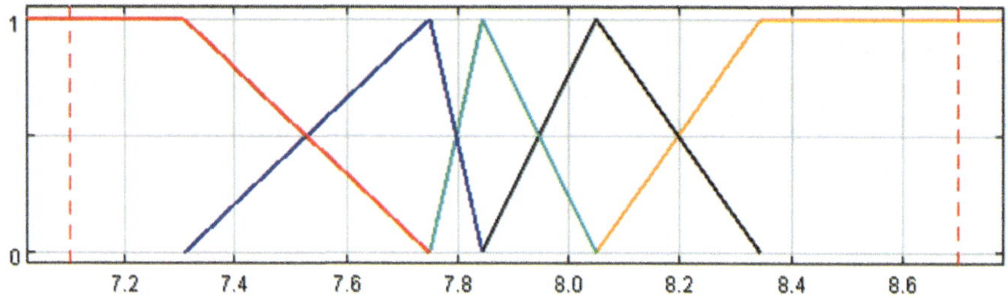

Fig. 10 Membership function of pH

After the analysis of the above experiments and their result analysis, the following findings are concluded.

1. HFP membership function partition with Wang–Mendel rule generation method has given more accurate results with a competent interpretability of the system.
2. In case of k-means MF generation, Wang–Mendel method has given more accurate result. However, the fast prototyping algorithm has performed poorly with a less accurate result.

5 Conclusion and Future Scope

The decision making of water quality has lot of subjectivity and needs a competent model to implement it. The application of computational intelligence techniques is a good idea to develop models for water quality assessment. In this paper, a fuzzy model is proposed to implement the policy of water quality assessment of river Ganga. Four important parameters are used in this decision making. Wang–Mendel method is found suitable for the computation with more accurate results.

In the future, the author is interested to develop the fuzzy water quality assessment models using type-2 and interval type-2 fuzzy models. It is also proposed to implement such models using Internet of things (IoT) techniques at the ground reality.

References

1. F.B. Semiromi, A.H. Hassani, A. Torabian, A.R. Karbassi, H. Lotfi, Water quality index development using fuzzy logic: a case study of the Karoon River of Iran. Afr. J. Biotechnol. **1**(50), 10125–10133 (2011)
2. A. Lermontov, L. Yokoyama, M. Lermontov, M.A.S. Machado, River quality analysis using fuzzy water quality index: Ribeirado Iguape river water shed. Braz. Ecol. Indic. **9**(6), 1188–1197 (2009)
3. Y. Lcaga, Fuzzy evaluation of water quality classification. Ecol. Ind. **7**(3), 710–718 (2007)
4. R. Li, Z. Zou, A. An, Water quality assessment an in on river based on fuzzy water pollution index method. J. Environ. Sci. **50**, 87–92 (2016)
5. H. Gharibi, A.H. Mahvi, R. Nabizadeh, H. Arabalibeik, M. Yunesian, M.H. Sowlat, A novel approach in water quality assessment based on fuzzy logic. J. Environ. Manag. **112**, 87–95 (2012)
6. W. Ocampo-Duque, N. Ferre-Huguet, J.L. Domingo, M. Schuhmacher, Assessing water quality in rivers with fuzzy inference systems: a case study. Environ. Int. **32**(6), 733–742 (2006)
7. S. Dahiya, B. Singh, S. Gaur, V.K. Garg, H.S. Kushwaha, Analysis of ground water quality using fuzzy synthetic evaluation. J. Hazard. Mater. **147**(3), 938–946 (2007)
8. N.-B. Chang, H.W. Chen, S.K. Ning, Identification of river water quality using the fuzzy synthetic evaluation approach. J. Environ. Manage. **63**(3), 293–305 (2001)
9. F. Nasiri, I. Maqsood, G. Huang, N. Fuller, Water quality index: a fuzzy river pollution decision support expert system. J. Water Resour. Plan. Manag. **133**(2) (2007)
10. S.K. Kumar, R. Bharani, N.S. Magesh, P.S. Godson, N. Chnadrasekar, Hydrogeochemistry and ground water quality appraisal of part of south Chennai wastal aquifers, Tamilnadu, India using WQI and fuzzy logic method. Appl. Water Sci. **4**(4), 341–350 (2014)
11. V.R. Raman, R. Bouwmeester, S. Mohan, Fuzzy logic water quality index and importance of water quality parameters, Air. Soil Water Res. **2**, 51–59 (2009)
12. H. Gharibi, M.H. Sowlat, A.H. Mahvi, H. Mahmoudzadeh, H. Arabalibeik, M. Keshararz, N. Karimzadeh, G. Hassani, Development of dairy cattle drinking water quality index (DCWQI) based on fuzzy inference systems. Ecol. Ind. **20**, 228–237 (2012)
13. P. Abdullah, S. Waseem, B.V. Raman, I.-U. Mohsin, Development of a new water quality model using fuzzy logic system for Malaysia. Open Environ. Sci. **2**, 101–106 (2008)
14. A. Mourhir, T. Rachidi, M. Karim, River water quality index for Morocco using a fuzzy inference system. Environ. Syst. Res. **3**(21) (2014)
15. J.J. Carbajal-Hernandez, L.P. Sánchez-Fernandez, J.A. Carrasco-Ochoa, J.F. Martínez-Trinidad, Immediate water quality assessment in shrimp culture using fuzzy inference system. Expert Syst. Appl. **39**(12), 10571–10582 (2012)
16. D. Wang, V.P. Singh, Y. Zhu, Hybrid fuzzy and optimal modeling for water quality evaluation. Water Resour. Res. **43**, 1–10 (2007)
17. S.R.M.M. Roveda, A.P.M. Bondanca, J.G.-S. Silva, J.A.F. Roveda, A.H. Rosa, Development of a water quality index using a fuzzy logic: a case study for the Sorocaba river, in *International Conference on Fuzzy Systems*, Barcelona, Spain (2010), pp. 18–23
18. S.S. Mahapatra, S.K. Nanda, B.K. Panigrahy, A cascaded fuzzy inference system for Indian river water quality prediction. Adv. Eng. Softw. **42**(10), 787–796 (2011)
19. A.M. Jinturkar, S.S. Deshmukh, S.V. Agarkar, G.R. Chavhan, Determination of water quality index by fuzzy logic approach: a case of ground water in an Indian Town. World Sci. Technol. **61**(8), 1987–1994 (2010)
20. D. Scannapieco, V. Naddeo, T. Zarra, V. Belgiorno, River water quality assessment: a comparison of binary and fuzzy logic based approaches. Ecol. Eng. **47**, 132–140 (2012)
21. P.K. Shukla, S.P. Tripathi, A review on the interpretability-accuracy trade-off in evolutionary multi-objective fuzzy systems (EMOFS). Inf. Sci. **3**(3), 256–277
22. P.K. Shukla, S.P. Tripathi, A survey on interpretability-accuracy trade-off in evolutionary fuzzy systems, in *Fifth International Conference on Genetic and Evolutionary Computing (ICGEC)*, 29 August–01 September (2011), pp. 97–101

23. P.K. Shukla, S.P. Tripathi, Interpretability issues in evolutionary multi-objective fuzzy knowledge base system, in *Seventh International Conference on Bio-Inspired Computing: Theories and Applications (BIC-TA 2012)*, ABV-IIITM, Gwalior (2012), pp. 473–484
24. P.K. Shukla, S.P. Tripathi, Handling high dimensionality and interpretability-accuracy trade-off issues in evolutionary multi-objective fuzzy classifiers. Int. J. Sci. Eng. Res. **5**(6), 665–670 (2014)
25. S.P. Tripathi, P.K. Shukla, Uncertainty handling using fuzzy logic in rule based systems. Int. J. Adv. Sci. Technol. **45**, 31–46 (2012)
26. P.K. Shukla, S.P. Tripathi, On the design of interpretable evolutionary fuzzy systems (I-EFS) with improved accuracy, in *International Conference on Communication Systems*, Sept. 14–Sept. 15, Phagwara, India (2012), pp. 11–14
27. P.K. Shukla, S.P. Tripathi, Interpretability and accuracy issues in evolutionary multi-objective fuzzy classifiers. Int. J. Soft Comput. Netw. **1**(1), 55–69 (2016)
28. P. Chandra, D. Agarawal, P.K. Shukla, MOBI-CLASS: a fuzzy knowledge based system for mobile handset classification, in *7th International Conference on Soft Computing for Problem Solving (SocProS),* IIT Bhubaneswar, India, Dec. 23–24 (2017)
29. https://nmcg.nic.in/NamamiGanga.aspx. Accessed 01 March 2019
30. J.M. Alonso, L. Magdalena, HILK++: an interpretability-guided fuzzy modeling methodology for learning readable and comprehensible fuzzy rule-based classifiers. Soft. Comput. **15**(10), 1959–1980 (2011)

A Hybrid Framework for Fire Outbreak Detection Based on Interval Type-2 Fuzzy Logic and Flower Pollination Algorithm

Uduak A. Umoh, Udoinyang G. Inyang, and Emmanuel E. Nyoho

Abstract In this paper, a hybrid approach for fire outbreak detection based on interval type-2 fuzzy logic (IT2FL) and flower pollination algorithm (FPA), using environmental parameters is proposed. Due to the high uncertainty in the fire outbreak data, IT2FLSs are able to consider many linguistic uncertainties in the membership functions (MFs) of the type-2 framework, thus, it can raise the accuracy of the fuzzy system. The MF parameters of IT2FL controller are optimized by the flower pollination algorithm (FPA). The fire outbreak data capturing device (FODCD) is developed to extract fire outbreak environmental data and store in a database. The proposed optimized controller is simulated and experimentally applied to detect a fire outbreak. The controller performance is compared with the conventional type-2 fuzzy logic-based controller, respectively, in the MATLAB/Simulink environment. The experimental result indicates that with the temperature at 40.657 °C, smoke at 77.86%, flame at 762.95 ppm (part per million) and T (threshold) of 0.8, the IT2FL-FPA and IT2FL predict fire outbreak with 0.8276642 (83%) and 0.777972 (78%) possibility, respectively. The performance results show that when the threshold T is kept between an optimal range of 0.8 and 0.85, the IT2FL-FPA model gives accuracy between 93.33% and 100% with an error rate between 0 and 0.07%, while the IT2FL model gives accuracy between 90 and 96.67% with an error rate between 0.03 and 0.1%. The simulation and experimental results show that the IT2FL-PFA controller outperforms the same controller without optimization.

Keywords Flower pollination algorithm · Interval type-2 fuzzy system · Linguistic uncertainties · Optimized

---

U. A. Umoh (✉) · U. G. Inyang · E. E. Nyoho
Department of Computer Science, University of Uyo, PMB 1017, Uyo, Akwa Ibom State, Nigeria
e-mail: uduakumoh@uniuyo.edu.ng

© Springer Nature Singapore Pte Ltd. 2020

A. K. Nagar et al. (eds.), *Soft Computing for Problem Solving 2019*,
Advances in Intelligent Systems and Computing 1139,
https://doi.org/10.1007/978-981-15-3287-0_3

1 Introduction

Fuzzy control, a method based on fuzzy logic [1], has gained very much interest in recent years. The increased popularity of fuzzy control can be attributed to the fact that fuzzy logic systems (FLS) provide a powerful vehicle that allows engineers to incorporate human reasoning in the control algorithm. Fuzzy control is a control method based on fuzzy logic [2, 3]. The objective of the FLS is to map the crisp input to the crisp output [4]. The fuzzy controllers designed using fuzzy logic implements human reasoning that has been programmed into membership functions (MF), fuzzy rules and rule interpretation. For an optimal result, fuzzy logic parameters are normally optimized with algorithms such as flower pollination algorithm (FPA), particle swarm algorithm (PSO) and genetic algorithm (GA).

FPA was developed by [5] as a nature-inspired metaheuristic optimization algorithm that imitates the pollination behavior of flowers. The main objective of FPA is to produce an optimal reproduction of plants for surviving the fittest flowers in the flowering plant. FPA process is the transfer of pollen by using pollinators such as insects, birds and bats. There are two major processes for transferring the pollens: biotic (cross-pollination) and abiotic (self-pollination). Fuzzy controllers, when optimized with the FPA algorithm, are able to make effective decisions on the basis of linguistic information. One of the applications of the fuzzy controller is in the area of fire detection.

Fire is a combustion process which results in light, smoke, heat, flame and various hazardous gases [6]. Heat, oxygen and fuel are the three major elements of fire. The nature of the fire depends on the proportion of each of these elements. Even though the fire has always been useful for promoting the development of human society and it is extensively used in a lot of applications such as cooking, manufacturing process and other processes, when fire is out of control, i.e., fire hazard, it can cause a serious effect to human life as well as property. Fire hazards can also damage the ecological environment on a great level [7].

Fire outbreak detection is the process of sensing of one or more phenomena resulting from the fire such as smoke, heat, infrared light radiation or gas. Traditionally the fire detection is done using only smoke detectors which is less reliable technique and results in false alarms, and these fire detectors are unable to respond quickly and in a reliable manner in hazardous conditions [8]. Fuzzy logic is applied in fire outbreak detection and monitoring [9, 10]. In this paper, a framework for fire outbreak detection based on IT2FL and FPA is proposed. Unlike the traditional fire outbreak detection devices, this system implements a sensor approach in predicting the fire outbreak.

2 Related Literature

Souissi and Ammar [24] worked on a new approach based on wireless sensor network and fuzzy logic for forest fire detection. The proposed system consists of three sensors—temperature sensor, humidity sensor and wind sensor. The Author used fuzzy logic algorithm implemented with MATLAB/Simulink to simulate the system. The system was able to detect fire risk levels during summer, winter, autumn and spring.

Karakose and Akin [18] worked on a block-based fuzzy controller for speed control of induction motor. He proposed a rule-based reduction method for Mamdani type fuzzy controllers using the MATLAB fuzzy logic toolbox. The performance of the system shows that the computational time of the conventional fuzzy controller can be reduced by 50% using the block-based fuzzy controller.

Khule and Jangle [14] designed and implemented a fire and obstacle detection and control system based on Arduino Uno using fire sensor, temperature sensor, smoke sensor and PIR motion sensor on automobile. The type-1 fuzzy logic system was developed with MATLAB. The microcontroller unit responds to the instructions sent by the MATLAB R2014 software according to the necessity of the application as well as triggers the relay and fan upon critical situations.

Chu [4] designed a fire detection algorithm based on fuzzy logic, using temperature sensor to estimate the direction of fire disaster. The sensor nodes used has the ability to periodically collect ambient temperature, communicate with neighbor sensors and store the neighbors' information such as nodes' ID and the nodes' coordinate.

Lucian et al. [20] on "Efficient Increase for Electrical Fire Detection and Alarm Systems through Implementation of Fuzzy Expert Systems," proposed a fuzzy logic expert system based on differential temperature (Td), temperature (Ta), smoke density variation (Sd), smoke density (Sa) and concentration of carbon monoxide (CO). Fuzzy logic toolbox and Simulink tool from MATLAB were used in the simulation of the system. The fuzzy expert system showed faster fire detection rate and a better resilient to transient phenomenon responsible for false alarms production.

Khanna and Cheema [19] on "Fire Detection Mechanism using Fuzzy Logic," implement fuzzy logic on the information collected by sensors. They used multiple sensors for detecting the probability of fire as well as direction of fire. Each sensor node consists of multiple sensors that will sense temperature, humidity, light and CO density for calculating the probability of fire and azimuth angle for calculating the direction of fire. The proposed forest fire detection handles the uncertainty present in the data effectively and gives the best results with very low false alarm rate.

Nunes et al. [22] on "Optimized Flame Detection," builds a model using fuzzy logic or fuzzy inference system (FIS) to detect fire pixels in an image. They focused on optimizing the flame detection by identifying gray cycle pixels nearby the flame. These techniques can be used to reduce false alarm by giving the accurate result of fire occurrence.

Necsulescu and Xuqing [21] on "Type-2 Fuzzy Logic Sensor Fusion for Fire Detection Robots," presented an approach for fire detection and estimation robots. The approach is based on type-2 fuzzy logic system that utilizes measured temperature and light intensity to detect fires of various intensities at different distances. The simulation work is achieved by using MATLAB simulation based on the proposed type-2 fuzzy logic sensor fusion approach. The type-2 fuzzy logic system (T2 FLS) is known for not needing the exact mathematical model and for its capability to handle more complicated uncertain situations compared to type-1 fuzzy logic system (T1 FLS).

Rahimi et al. [23] on "Integrated Fuzzy Control of Temperature, Light and Emergency Conditions for Smart Home Application," presented an integrated fuzzy controller in order to control light, temperature and emergency conditions. They presented the structure of integrated controller that is able to monitor and control emergency conditions, lighting and heating processes simultaneously. Heating, lighting, blind and fire-alarms are simulated with MATLAB/Simulink and the relationships between each of these systems are intended as well. The presented approach shows that designing a controller with fuzzy logic is simpler and also has a faster response than PID controllers.

3 The Proposed Framework

Figure 1 gives a hybrid framework for fire outbreak detection based on IT2FL and FPA to monitors fire outbreak situation via the use of temperature sensor, smoke sensor and flame sensor attached to an Arduino nano microcontroller unit.

The system collects the data from these sensors and transmits them via an ESP8266 Wi-Fi module to an IT2FL controller module. The collected data (input parameters) are then fuzzified (i.e., converted to an IT2F set). This fuzzy set is passed to an inference engine which evaluates the fuzzy set against the rules in the rule base and produces another T2F set. The fuzzy set produced during inference is reduced to type-1 fuzzy set using Karnik-Mendel algorithm. The reduced set is then defuzzified to give a crisp output. In the end, the input parameters together with the crisp output are stored in a fire dataset for future use.

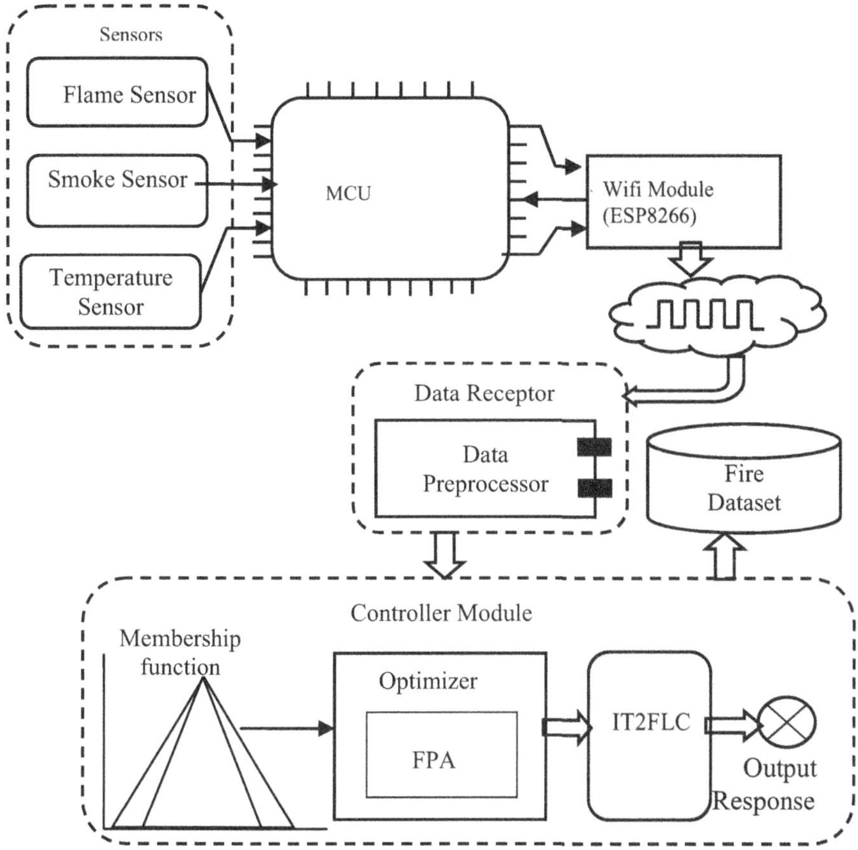

Fig. 1 Proposed system framework for fire outbreak detection

The various components of this system are; sensors—DHT11 temperature sensor, MQ-2 smoke sensor and LM393 flame sensor are used as input parameters to measure environmental parameters related to fire outbreak. Microcontroller unit (MCU)—The Arduino Nano V3.0 microchip is used to interface all the system peripherals (sensors and Wi-Fi module). The MCU serves as the brain of the data capture device to read parameter values from the connected sensors and send the same to the FPA-fuzzy module via an ESP8266 Wi-Fi module. Data receptor—The data receptor is implemented as a software module. It sends HTTP request for data to the ESP8266 Wi-Fi module and accepts a response in the form of "HTTP Response Text." The format "HTTP Response Text" is presented below;

```
<html>
   <head><title>FODCD Data</title></head>
   <body>
      *temp,smoke,flame#
   </body>
</html>
```

The Data Receptor reads and extracts the relevant parameters (i.e temperature, smoke, and flame values) from the "HTTP Response Text" using the following heuristic algorithm;

```
algorithm Preprocess(msg):Arr[]
begin:
   Define temp, smoke, flame;
   Define dataSize N;
   Define indexOfAsteric indx1, indexOfAsh indx2;
   Define substringVariablesubStr;
   Define array Variable Arr[];
   //find the position of char *
   Loop for i = 0 to N
      Let x = charAt(i)
      If x == "*"
         indx1 = i;
         break;
      end if
   end for
   //find the index of char #
   Loop for j = 0 to N
      Let y = charAt(j)
      if y == "#"
         Indx2 = j;
         break;
      end if
   end for
   //extract data substring and split to an array
   if( indx1 > -1 AND indx2 > -1)
      subStr = subString(indx1, indx2);
      Arr[] = split(subStr, ",");
      return Arr[];
   end if
end
```

Controller module—The controller module is made up of the FPA optimizer and the IT2FLC [11]. While FPA is used to tune the parameter values of the MFs, ITFLC maps the crisp input received from the three sensors into a defuzzified value used in predicting the occurrence of fire outbreak situation. The controller module is implemented as an android application.

4 The Flower Pollination Algorithm (FPA)

FPA is based on the three rules;

Rule 1: Biotic (cross-pollination) can be considered the process of global pollination, and pollen carrying pollinators move in a way that obeys Levy flights.
Rule 2: Abiotic (self-pollination) are considered processes of local pollination.
Rule 3: Pollinators such as insect can develop flower constancy, which is equivalent to a reproduction probability that is proportional to the similarity of two flowers involved.
Rule 4: The interaction or switching of local pollination and global pollination can be controlled by a switch probability $P \in [0, 1]$, slightly biased toward local pollination.

Mathematically, Rule 1 called global pollination and flower constancy can be represented as:

$$x^{t+1} = x_i^t + \gamma L(\lambda)\left(g_* - x_i^t\right) \quad (1)$$

$\Gamma(\lambda)$ is a gamma standard function and this distribution applies for large steps $s > 0$. Where: x^{t+1}: is the new flower or pollen gamete (candidate solution). x_i^t: is the current flower or pollen gamete. γ: is the scaling factor which controls the step size of each iteration; $L(\lambda)$: is the Levy flight distribution, the strenght of pollination, which is essentially a step size; g *: is the best current solution among all solutions. The movement of pollens (candidate solution) in flower pollination algorithm follows Levy flight distribution presented in Eq. (2);

$$L \sim \frac{\lambda\Gamma(\lambda)\sin\left(\frac{\pi\lambda}{2}\right)}{\pi}\frac{1}{s^{1+\lambda}}, (s \gg s_0 > 0) \quad (2)$$

Local pollination, which mimics the movement of pollen from one flower to another flower of the same plant, is modeled by the equation presented in Eq. (3);

$$x^{t+1} = x_i^t + \epsilon\left(x_j^t - x_k^t\right) \quad (3)$$

where x^{t+1} is the pollen from different flowers of the same plant. Mathematically, if x_j^t and x_k^t are pollens of the same plant or selected from the same population, this equivalently becomes a local random walk if we draw ϵ from a uniform distribution in [0, 1], i.e., $\epsilon \sim [0, 1]$. The switch probability form Rule 4 is used to effectively switch between global pollination to local pollination. The switching probability is presented mathematically as;

$$p \in [0, 1] \quad (4)$$

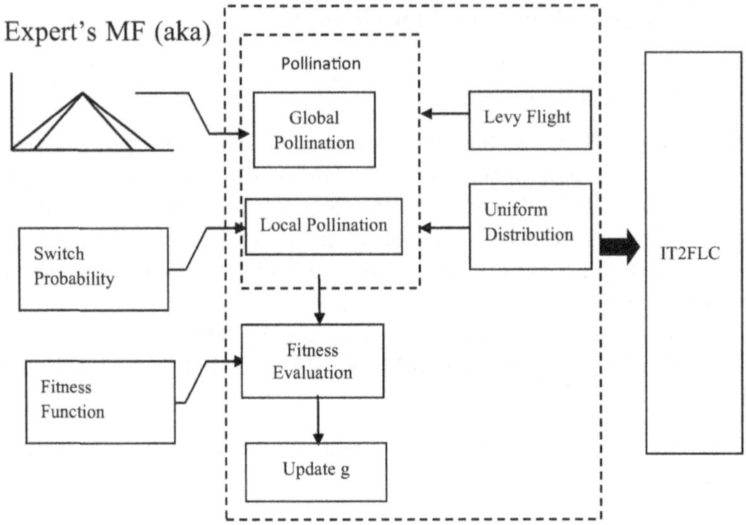

Fig. 2 Flower pollination algorithm

The FPA is implemented using the following pseudocode; *Step* 1: initialize a population of n flowers or pollen gametes with random solutions. *Step* 2: Find the best solution g_* in the initial population, define a switch probability; $p \in [0, 1]$. *Step* 3: while (t < maxGeneration) For i = 1: n; If rand < p; draw a (d-dimensional) step vector from a Levy distribution. Do global pollination via $\boldsymbol{x}^{t+1} = \boldsymbol{x}_i^t + \gamma \boldsymbol{L}(\boldsymbol{\lambda})\left(\boldsymbol{g}_* - \boldsymbol{x}_i^t\right)$ Else, draw from a uniform distribution [0, 1]; do local pollination via $\boldsymbol{x}^{t+1} = \boldsymbol{x}_i^t + \epsilon\left(\boldsymbol{x}_j^t - \boldsymbol{x}_k^t\right)$; End if; Evaluate new solution. If a new solution is better, update them in the population. End for; Find the current best solution $\boldsymbol{g}_*$; End while. *Step 4:* Output the best solution found. The FPA is represented graphically in Fig. 2.

Real-valued encoding is used to convert the MFs from the problem space to FPA candidate solution called flower or pollen gamete. The dimension of the problem d is 26, calculated as;

$$d = \left[\sum_{i}^{V} \sum_{j}^{T} P_{ij} \right] \tag{5}$$

where: V—is the total number of linguistic variables (3 in this case). T—is the total number of linguistic terms of the ith variable. Pij—is the number of an adjustable parameter(s) of linguistic term j in variable i. The objective function used in this paper is adapted from [12] and presented in Eq. (6);

$$f_i = e^{3(n-i)/n} \tag{6}$$

where: e—is the exponential function; n—total number of iteration; i—current iteration.

5 Interval Type-2 Fuzzy Logic for Fire Outbreak Detection

The paper adapts and modifies the IT2FLC controller in [13, 14] and presented in Fig. 3. Fuzzification module maps the crisp input (fire outbreak parameter values) to an IT2 set using Gaussian MF. Inference engine (IE) module evaluates the rules in the knowledge base against IT2F set from the fuzzification module, to produce another IT2F set. Type reducer (TR) uses K-M algorithm to reduce an IT2F set to T1F set. Defuzzification module maps the fuzzy set produced by TR to a crisp output using the center of gravity method.

Knowledge base is a database of rules (rules are generated from experts' knowledge) to be used by the inference engine. MF helps the fuzzification module convert the crisp input into a fuzzy set [11].

The proposed framework is based on the three linguistic variables (parameters)—Temperature, smoke and flame. The linguistic variables and their universe of discourse are presented in Table 1.

The Gaussian MF is presented in Eqs. (7–9);

$$f(x) = e^{\frac{-1}{2}\left(\frac{x-c}{\sigma}\right)^2}, \sigma \in [\sigma_1, \sigma_2] \text{ and } c \in [c_1, c_2] \tag{7}$$

$$\underline{\mu}_{\tilde{A}_{im}}(x_i) = e^{\frac{-1}{2}\left(\frac{x_i - c_{im}}{\underline{\sigma}_{1,im}}\right)^2}, \underline{\mu}_{\tilde{A}}(x) = N(c, \sigma_1; x) \tag{8}$$

$$\overline{\mu}_{\tilde{A}_{im}}(x_i) = e^{\frac{-1}{2}\left(\frac{x_i - c_{im}}{\overline{\sigma}_{2,im}}\right)^2}, \overline{\mu}_{\tilde{A}}(x) = N(c, \sigma_2; x) \tag{9}$$

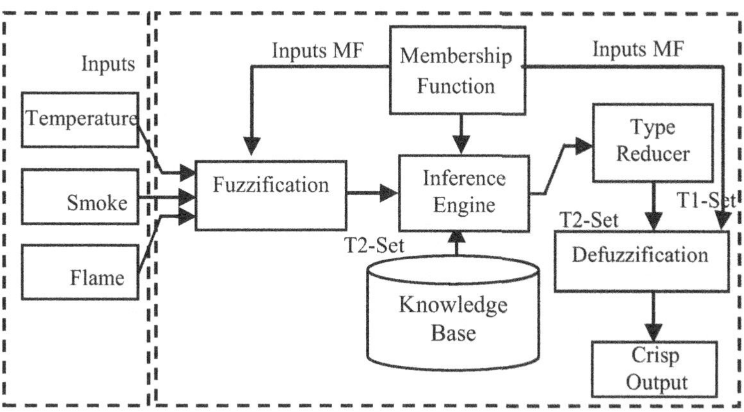

Fig. 3 IT2FL model for fire outbreak detection

Table 1 Universe of discourse

Linguistic variable	Universe of discourse	Description
Temp	[0, 50]	Temperature (input)
Smoke	[0, 100]	Smoke (input variable)
Flame	[0, 1023]	Flame (input variable)
Fire outbreak	[0, 1]	Output variable

where c_1, c_2—are the centers of the lower and upper Gaussian MFs, respectively. σ_1, σ_2—is the variance (or width) of the lower and upper MFs, *e*—is the exponential function, x_i—is the input vector, $\overline{\mu}_{\tilde{A}_{im}}(x_i)$—is the degree of membership of input x_i in the variable, *i* of linguistic term, *m*. The optimized membership plots of this system are presented in Figs. 4, 5 and 6. The output MF—"Fire Outbreak" is presented in Fig. 7.

The IT2FL rules are defined based on Eq. (10) [15].

$$R^l : IF\, x_1\, is\, \widetilde{F}_1^l\, and \ldots x_p\, is\, \widetilde{F}_p^l\, THEN\, y\, is\, \widetilde{G}_1^l \quad (10)$$

where l is the rule number $l = 1, \ldots, M$; x_1 : is the linguistic variable, $\widetilde{F}^l$: is the linguistic term. For example, IF temp is Low AND smoke is Low AND flame is VeryLow THEN FireOutbreak is VeryLow. Where temp, smoke and flame are the linguistic variables while Low and VeryLow are linguistic values (terms) determined

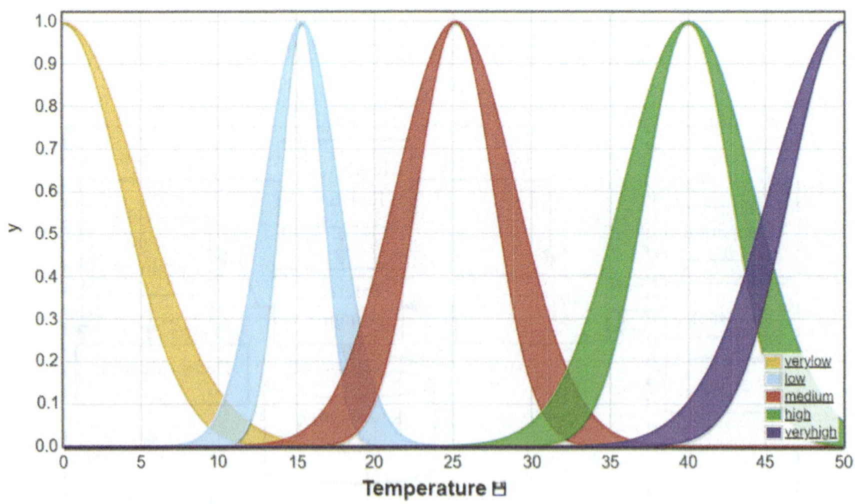

Fig. 4 Temperature membership function

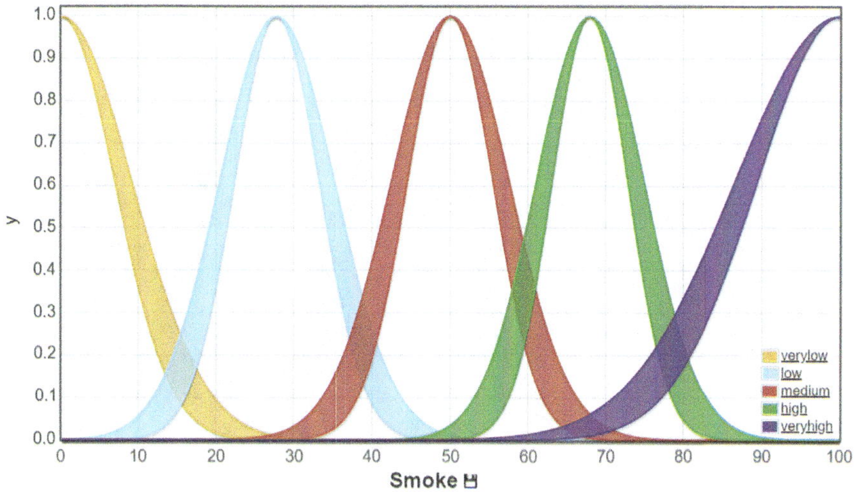

Fig. 5 Smoke membership function

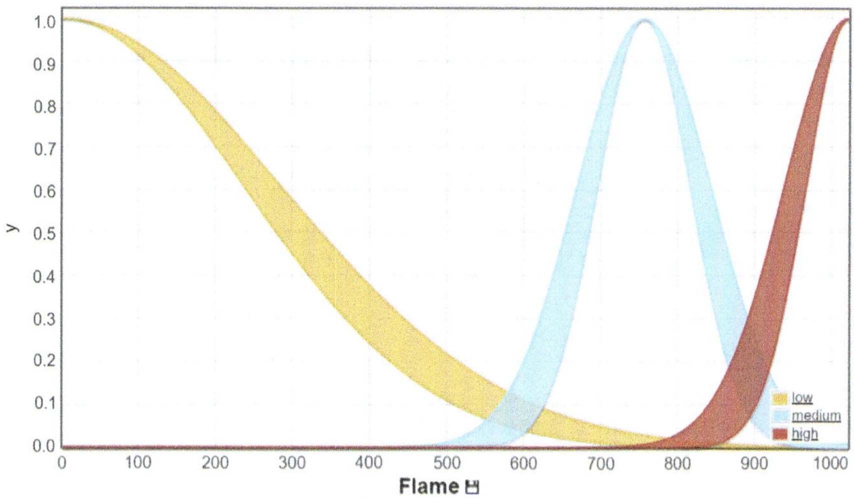

Fig. 6 Flame membership function

by fuzzy sets on the universe of discourse of temp, smoke, flame and fire outbreak, respectively. The centroid of the consequent set $[\underline{y}^1, \bar{y}^1]$, is calculated using;

$$\underline{\mu}_{\overline{B}}(y) = \underline{\mu}_F(x) * \underline{\mu}_G(y) \quad \forall y \epsilon Y_d \tag{11}$$

$$\overline{\mu}_{\overline{B}}(y) = \overline{\mu}_F(x) * \overline{\mu}_G(y) \quad \forall y \epsilon Y_d \tag{12}$$

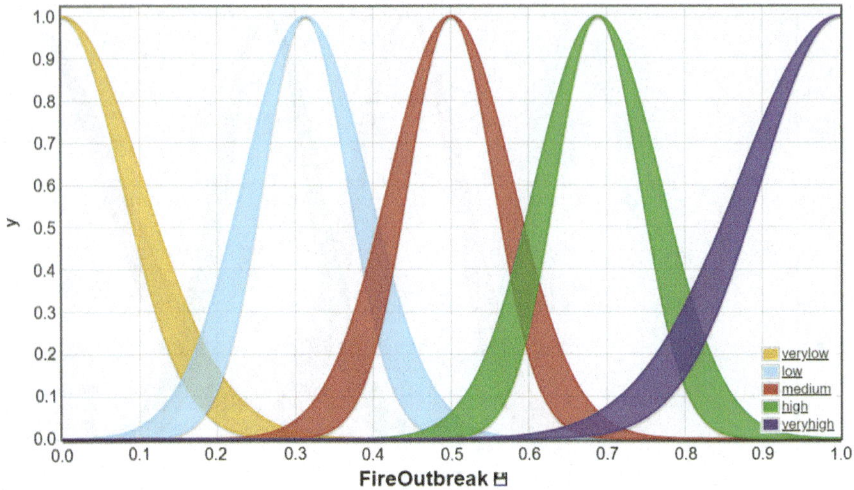

Fig. 7 Membership function for fire outbreak

Table 2 presents the subset of the fuzzy rules used in this system;

A Mamdani type inference mechanism [16] is used to evaluate the fuzzy rules against the fuzzy set calculated from their respective MFs is presented in Eqs. 13, 14 and 15.

$$F^i(x') = \left[\underline{f}^i(x'), \bar{f}^i(x')\right] \equiv \left[\underline{f}^i, \bar{f}^i\right] \tag{13}$$

$$\underline{f}^i(x') = \underline{\mu}_{\tilde{F}_1^i(x')} * \ldots * \underline{\mu}_{\tilde{F}_i^i(xp')} \tag{14}$$

$$\bar{f}^i(x') = \overline{\mu}_{\tilde{F}_1^i(x')} * \ldots * \overline{\mu}_{\tilde{F}_i^i(xp')} \tag{15}$$

Table 2 Fuzzy logic rules for fire outbreak detection

S/N	Temp	Smoke	Flame	Fire outbreak
1	VL	VL	L	VL
2	VL	VL	M	VL
3	VL	VL	H	VL
4	VL	L	L	VL
5	VL	L	M	VL
6	VL	L	H	L
7	VL	M	L	VL
8	VL	M	M	L
9	VL	M	H	L
10	VL	H	L	L

where; $\left[\underline{f}^i, \bar{f}^i\right]$—is the firing antecedent set. $\underline{\mu}_{\tilde{F}_1^i(x')}$—is the fuzzy set of the lower MF. $\bar{\mu}_{\tilde{F}_1^i(x')}$—is the fuzzy set of the upper MF. The IE combines the fired rules and gives a mapping from input IT2FSs to output IT2FSs.

The paper employs K-M TR algorithm [17] to reduce the IT2F set to a T1F set using defuzzification module by algorithm seeks to find switch points (L, R) and the leftmost point (y_l) and the rightmost point (y_r) determined by;

$$\underline{y}^L \leq y_l \leq \underline{y}^{L+1} \tag{16}$$

$$\bar{y}^R \leq y_r \leq \bar{y}^{R+1} \tag{17}$$

where $\underline{y}^n$ *and* $\bar{y}^n$ have been sorted in ascending order, we explore the center of sets type reducer (COS) method as summarized in Eqs. (18)–(21). The details descriptions are found in Eq. 4.

$$Y_{TR}(x') = [y_l(x'), y_r(x')] \equiv$$

$$[y_l, y_r] = \bigcup_{\substack{f^i \in F^i(x') \\ y^i \in Y^i}} \frac{\sum_{i=1}^N f^i y^i}{\sum_{i=1}^N f^i} \tag{18}$$

which,

$$y_r = \frac{\sum_{i=1}^N f_r^i y_r^i}{\sum_{i=1}^N f_r^i} \tag{19}$$

$$y_l = \frac{\sum_{i=1}^N f_l^i y_l^i}{\sum_{i=1}^N f_l^i} \tag{20}$$

Equation 21 computes the output as;

$$y = \frac{y_l + y_r}{2} \tag{21}$$

6 IT2FL-PFA Membership Function Encoding

Figure 1 is used to tune the MF parameters of the IT2FLC. Real-valued encoding is used to convert the MFs from the problem space to FPA candidate solution called flower of pollen gamete. The dimension of our problem *d* is 26 calculated as;

$$d = \left[\sum_{i}^{V} \sum_{j}^{T} P_{ij} \right] \tag{22}$$

where: V—is the total number of linguistic variables

T—is the total number of linguistic terms of the ith variable

P_{ij}—is the number of the adjustable parameter(s) of linguistic term j in variable i. FPA objective function as used in this paper defines how good a candidate solution is [10] and presented in Eq. (23);

$$f_i = e^{3(n-i)/n} \tag{23}$$

where *e*—is the exponential function; *n*—total number of iteration; *i*—the current iteration.

7 Results

Figure fire outbreak data capture device was used to capture our input parameter values (i.e., temperature values, smoke values and flame values). The 30 (thirty) data points obtained are presented in Table 3; these input parameters are fuzzified using an FPA tuned fuzzy logic membership functions whose training result is presented in Fig. 8.

Table 3 Obtained parameter values

S/N	Temp	Smoke	Flame
1	27.263	52.441	546.63
2	49.873	82.65	900.66
3	40.657	77.86	762.95
4	36.076	68.233	450.37
5	19.826	37.872	389.94
6	37.674	71.794	565.7
7	21.81	36.77	397.36
8	32.906	60.597	414.79
9	17.881	28.9	269.92
10	29.021	53.883	507.25
13	39.609	64.403	706.88
…	…	…	…
28	26.908	51.535	523.7
29	21.221	39.101	312.77
30	35.374	67.543	484.69

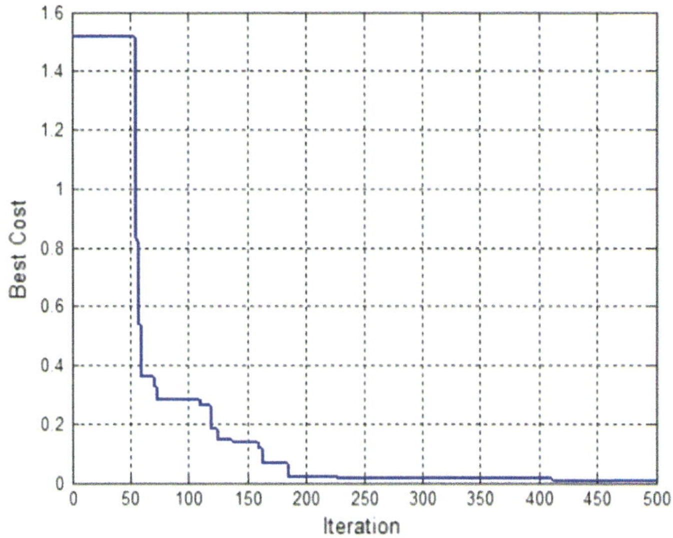

Fig. 8 FPA convergence graph for interval type-2 MFs tuning

The FPA convergence graph in Fig. 8 gives a minimal error (best cost) of 0.005923 which shows that the interval type-2 MF values are in its optimal configuration. The result of fuzzifying the input parameters is presented in Table 4.

Table 4 IT2FL-PFA fire outbreak prediction results

S/N	Temp	Smoke	Flame	IT2FL	IT2FL-FPA
1	27.263	52.441	546.63	0.45986	0.57639
2	49.873	82.65	900.66	0.83988	0.89073
3	40.657	77.86	762.95	0.77797	0.82766
4	36.076	68.233	450.37	0.76416	0.75727
5	19.826	37.872	389.94	0.38579	0.50588
6	37.674	71.794	565.7	0.74452	0.76684
7	21.81	36.77	397.36	0.32173	0.54469
8	32.906	60.597	414.79	0.57157	0.69267
9	17.881	28.9	269.92	0.30375	0.39827
10	29.021	53.883	507.25	0.49959	0.59547
11	18.115	32.369	287.57	0.32031	0.41129
12	17.591	29.446	228.11	0.29611	0.38363
15	35.558	58.21	567.57	0.68666	0.74033
….	…	…	…	…	…
29	21.221	39.101	312.77	0.37114	0.53705
30	35.374	67.543	484.69	0.73745	0.75419

The metrics used to evaluate the performance of this system include true positive (TP), false positive (FP), true negative (TN), false negative (FN), true positive rate (TPR), false positive rate (FPR), precision, accuracy and error rate. The performance evaluation of this system is presented in Table 5.

The graph of the results of IT2FL and IT2FL-FPA for fire outbreak detection showing the effect of the input parameters (temperature, smoke and flame) on the output (fire outbreak) as presented in Fig. 9; Table 6 shows the performance evaluation result of the effect of threshold on the system performance as visualized in Fig. 10 for IT2FL and IT2FL-FPA while Table 6 shows the performance evaluation result

Table 5 Performance evaluation

S/N	Threshold	TP	FP	FN	TN	TPR (recall)	FPR
1	0.6	3	10	0	17	1	0.3704
2	0.7	3	7	0	20	1	0.2593
3	0.8	3	0	0	27	1	0
4	0.85	1	0	2	27	0.333	0
5	0.9	0	0	3	27	0	0

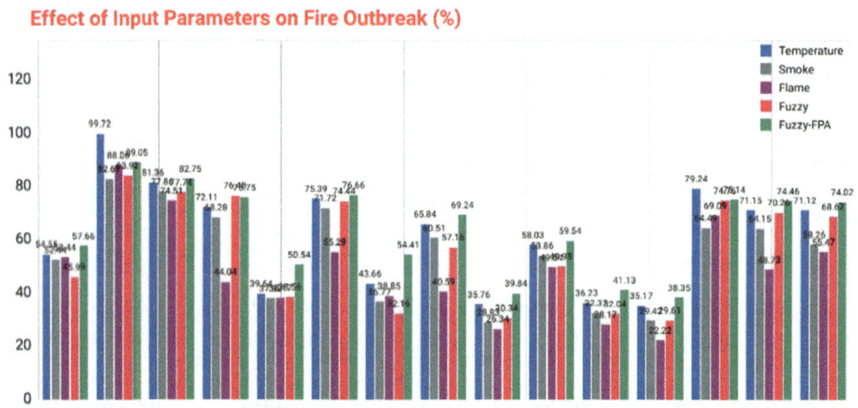

Fig. 9 Graph of IT2FL and IT2FL-FPA for fire outbreak detection results

Table 6 Performance evaluation

S/N	Precision	Accuracy (%)	Error rate
1	0.230769231	66.67	0.33
2	0.3	76.67	0.23
3	1	100	0
4	1	93.33	0.07
5	0	90	0.1

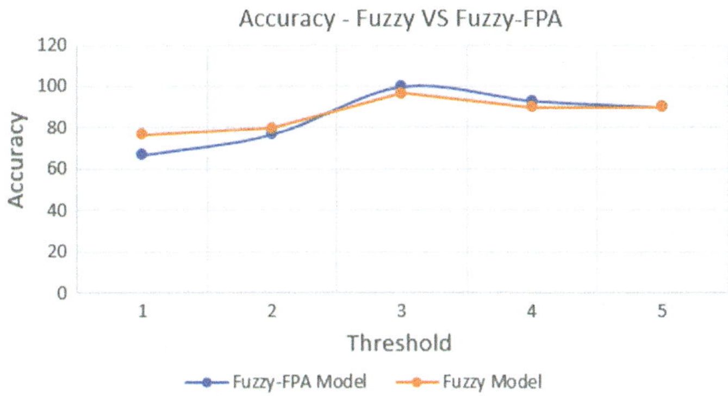

Fig. 10 Effect of threshold values on system accuracy for IT2FL and IT2FL-FPA

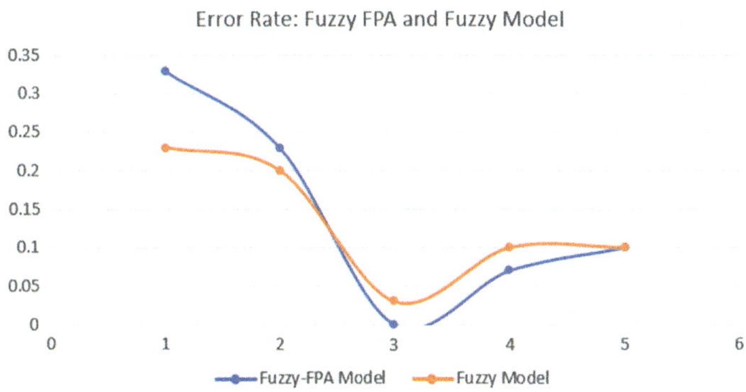

Fig. 11 Performance evaluation error rate for fuzzy and fuzzy-FPA models

based on precision, accuracy and error rate. Figure 11 gives the graph of performance evaluation error rate for fuzzy and fuzzy-FPA models.

8 Conclusion

This paper proposed a framework for fire outbreak detection based on IT2FL and FPA to promptly predict cases of fire outbreak. A framework has been able to proffer solution to the problems associated with fire outbreak by providing continuous monitoring of environmental changes that are responsible for fire outbreak such as temperature, smoke and flame. The system is also able to detect cases of fire outbreak and take remedial actions such as notifying the house owner and then sending a distress text to the fire service with location information. The system can predict a case of fire outbreak with accuracy between 93.33 and 100% when the threshold value is set between an optimal range of 0.8 and 0.85 with a minimal error rate between 0 and 0.07%. It is possible to detect and take remedial action during a fire outbreak.

On implementation, the system can protect lives and properties as well as reduce the rate of depression resulting from fire disaster. The combination of sensors, fuzzy controller and the android system gives a better result to the problem of fire management. In the future, the system can be integrated with the particle swarm optimization tool for better performance.

References

1. L.A. Zadeh, Fuzzy sets and systems. J. Informat. Control 338–353 (1965)
2. J. Jantzen, *Design of Fuzzy Controllers*. Denmark: the Technical University of Denmark, Department of Automation (1999)
3. T.J. Ross, Fuzzy *Logic with Engineering Applications*, 2nd edn. Wiley
4. X.J. Chu, *Design of Fire Detection Algorithm Based on Fuzzy Logic and using Wireless Sensors* (Ottawa-Carleton Institute of Electrical and Computer Engineering (OCIECE), Canada, 2010)
5. S.Y. Xin, Multi-objective flower algorithm for optimization. Proc. Comput. Sci. **18**, 155–173 (2012)
6. D. Sharma, Implementation of artificial neural fuzzy inference system in a real-time fire detection mechanism. Int. J. Comput. Appl. **146**, 31–36 (2016)
7. Saeed, F., Paul, A., Rehman, A.., Hong, W.H., Seo, H., IoT-Based Intelligent Modeling of Smart Home Environment for Fire Prevention and Safety. J. Sens. *Actuator Netw*., Vol. 7, no. 11, (2018)
8. M.A. Mobin, M. Rafi, M.N. Islam, An intelligent fire detection and mitigation system safe from fire. Int. J. Comput. Appl. **133**, 1–6 (2016)
9. R. Sowah, A review on forest fire detection techniques. Int. J. Distribut. Sensor Netw. 2–8 (2014)
10. B. Sarwar, I.S. Bajwa, S. Ramzan, B. Ramzan, M. Kausar, Design and application of fuzzy logic based fire monitoring and warning systems for smart buildings. Symmetry **10**(615), 1–24 (2018). https://doi.org/10.3390/sym10110615
11. J.M. Mendel, *Introduction to Ruled-Based Fuzzy Logic Systems* (University of South California, California, 1995)
12. W. Chen, Membership functions optimization of fuzzy control based on genetic algorithms, in *International Refrigeration and Air Conditioning Conference*, pp. 207–210. Jiaotong: Purdue University (1998)
13. Q. Liang, J.M. Mendel, Interval Type-2 fuzzy logic systems: theory and design. IEEE Trans. Fuzzy Syst. **8**, 535–550 (2000)
14. V. Khule, N.N. Jangle, Design and implementation of a fire and obstacle detection and control system using fuzzy logic with notification system to avoid automobile accidents, in *10th International Conference on Recent Trends in Engineering Science and Management*. India: Newton's School of Science and Technology, pp. 307–313 (2017)
15. J. Mendel, *Uncertain Rule-Based Fuzzy Logic Systems: Introduction and New Directions*, 1st ed. Prentice-Hall, Upper Saddle River, NJ, 2(1) (2001)
16. E.H. Mamdani, E. H., S. Assilian, An experiment in linguistic synthesis with a fuzzy logic controller. Int. J. Man Mach. Stud. 1–13 (1975)
17. Q. Liang, N.N. Karnik, Connection admission control in ATM networks using survey-based type-2 fuzzy logic systems. IEEE Trans Syst Man Cybernet Part C Appl Rev **30**(3), 329–334 (2000)
18. M. Karakose, E. Akin, Block based fuzzy controllers. Int. J. Res. Rev. Appl. Sci **3**(1), 100–110 (2010)

19. V. Khanna, R.K. Cheema, Fire detection mechanism using fuzzy logic. Int. J. Comput. Appl. **65**(3), 5–9 (2013)
20. H. Lucian, P. Ralu, C. Mircea, Efficiency increase for electrical fire detection and alarm systems through implementation of fuzzy expert systems. *U.P.B. Sci. Bull., Series C,* 75(1): 251–266 (2013)
21. D. Necsulescu, L. Xuqing, Type-2 sensor fusion for fire detection robots, in *Proceedings of the 2nd International Conference of Control, Dynamic Systems, and Robotics*. No. 187, Ottawa, Ontario, Canada, 1–1 (2015)
22. A. Nunes, L. Dias, M. Ugala, S. Pereira, Optimal flame detection. Int. J. Comput. Appl. **113**(15), 41–44 (2015)
23. A. Rahimi, F. Fardad, J. Shahram, Integrated fuzzy control of temperature, light and emergency conditions for smart home application. Int. J. Smart Electr. Eng. **5**(2), 93–99 (2016)
24. R. Souissi, M.B. Ammar, A new approach based in wireless sensor network and fuzzy logic for forest fire detection. Int. J. Comput. Appl. **89**(2), 48–55 (2014)

Using Convolutional Neural Networks to Predict Colon Cancer Patients Survival

Rawan Gedeon, Atulya K. Nagar, and Raouf Naguib

Abstract This paper aims to predict colon cancer patients' survival by using deep learning to extract prognostic biomarkers from haematoxylin- and eosin (HE)-stained tissue slides. A deep convolutional neural network is trained by transfer learning using 100,000 HE images achieving a nine-class accuracy >97%. This model is then used to segment digital whole slide images from a cohort of patients from the Cancer Genome Atlas (TCGA). The classification map produced is then used to quantify tumour–stroma ratio and tumour-infiltrating lymphocytes regions. These are then evaluated for their prognostic value for overall survival (OS) in a multivariate Cox proportional hazard model.

1 Introduction

Colon cancer is the fourth most common cancer in the UK, and despite significant advances in its treatment, the 10-year overall survival is around 57% in England and Wales [1]. Survival depends on the stage of the cancer when it is diagnosed.

Tumour–stroma ratio (TSR) has been reported to be an effective prognostic parameter in colon cancer [2]. TSR is an estimate for the amount of stroma present in the primary tumour in haematoxylin- and eosin-stained histological sections. High stroma content has been linked with poor prognosis. When scoring TSR, pathologists choose a field of vision with the highest percentage of stroma, and tumour cells must be present at all borders of the selected field. Therefore, in some studies, a pathologist extracts image regions that are then processed by the deep learning algorithm to score the TSR [3]. However, in this study, we are investigating the prognostic value

R. Gedeon (✉) · A. K. Nagar · R. Naguib
Liverpool Hope University, Liverpool, UK
e-mail: gedeonr@hope.ac.uk

A. K. Nagar
e-mail: nagara@hope.ac.uk

R. Naguib
e-mail: naguibr@hope.ac.uk

© Springer Nature Singapore Pte Ltd. 2020

A. K. Nagar et al. (eds.), *Soft Computing for Problem Solving 2019*,
Advances in Intelligent Systems and Computing 1139,
https://doi.org/10.1007/978-981-15-3287-0_4

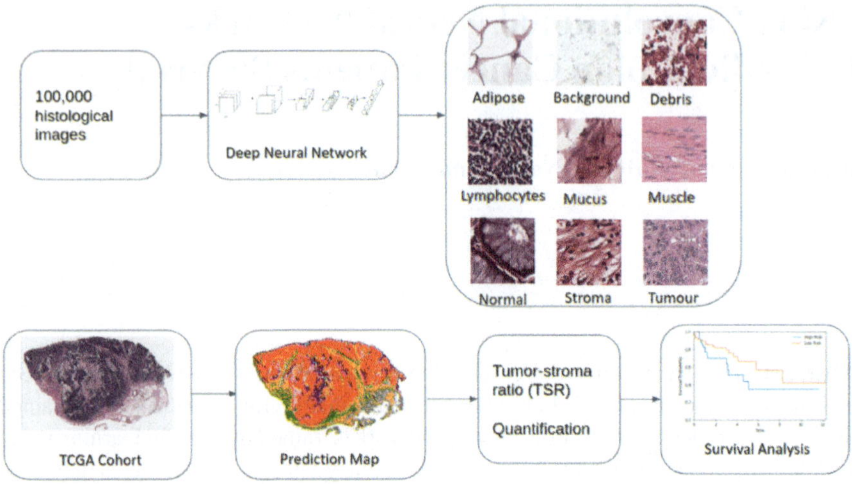

Fig. 1 Flowchart of the proposed approach

of TSR when calculated on the whole slide image level based on the tissue classification maps. We are also quantifying regions where both tumour and lymphocytes are co-localised. These regions are called tumour-infiltrating lymphocytes (TILs) to investigate their prognostic value. The digitisation of histology glass slides has made the analysis of the digitised whole slide images (WSIs) possible by image processing and deep learning algorithms.

The proposed approach is illustrated in Fig. 1. We first train a deep convolutional neural network (CNN) using 100,000 histological images of colorectal cancer and healthy tissues. Then, the CNN is used to segment the tissue regions in a WSI by classifying 50% overlapping image patches grouped together in a sliding window approach. Then, we quantify TSR and TILAb score and finally use them to predict patients' overall survival (OS) by employing univariate and multivariate analysis.

2 Materials and Methods

2.1 Data Set and Patient Cohorts

First, 100,000 histological images that were created from 86 HE-stained human cancer tissue slides were used to train a tissue image classifier. The data set [4] is from the NCT biobank and the UMM pathology archive published by a group of researchers from University of Heidelberg in Germany. We used the datatset that was not colour normalised. It consists of non-overlapping image patches from haematoxylin- and eosin (H&E)-stained histological images. All images are 224×224 pixels (px) at

Table 1 Summary of clinical parameters for TCGA COAD cohort

Clinical parameters		Full cohort	Modelling set	Test set
No. of patients		368	184	184
Age (years)		69	68.5	69
Overall survival (days)		390	362	414
Gender	Male	195	90	105
	Female	173	78	95
Tumour Stage	Stage I	64	35	29
	Stage II	150	78	72
	Stage III	94	57	37
	Stage IV	51	30	21
Patient status	Alive	283	141	142
	Dead	85	42	43

0.5 microns per pixel (MPP). Tissue classes are adipose, background, debris, lymphocytes, mucus, smooth muscle, normal colon mucosa, cancer-associated stroma, colorectal adenocarcinoma epithelium.

Second, we retrieved the HE slides for colon cancer patients from the TCGA cohort COAD with clinical data. After removing slides with artefacts such as tissue folding or marker scribbles, the cohort consisted of 368 diagnostic slides, one WSI for each patient. This TCGA data was used to analyse prognostic impact of neural network-based tissue decomposition with the end goal being predicting patients OS. A clinical information summary is provided in Table 1. Survival time is available from the date of the diagnosis. The median age for patients is 69 years. Approximately, 53% of the patients are male, while 58% are at stage I/II and remaining are at stage III/IV.

2.2 *Training Neural Networks*

We trained five of the most popular CNN models: Resnet [5], VGG [6], DenseNet [7], SqueeseNet [8] and Inception v3 [9] using transfer learning. The classification layer was replaced with a nine-class output layer. We divided the data set into 70% training set, 15% validation set and 15% testing set and trained all networks with stochastic gradient descent with a learning rate of 0.005, batch size of 360 and for 8 epochs. The code was implemented in PyTorch and ran on two NVIDIA RTX 24GB GPU. We used data augmentation with random horizontal and vertical flips for the training data.

2.3 Whole Slide Image Analysis

Whole slide images (WSI) are the digital replica of the glass slide where tissue sections had been stained with haematoxylin and eosin. They provide a very high resolution of the tissue structures and thus are very big in size, sometimes over 10GB. Therefore, they are not fed into a machine learning algorithm. Scaling them down would encompass a big loss of information, while cropping one region would not provide the whole picture for the analysis. In this study, we process all of the WSI by extracting patches in a sliding window approach. As a first step, we extract a tissue mask, as there is no need to process patches that do not contain any tissues. Tissue samples in this cohort were provided by multiple clinical centres. Therefore, they were captured differently; some WSI were digitised at 40X magnification and others at 20X. Therefore, careful consideration was taken when extracting patches as the classifier was trained using images extracted at 0.5 μm/px. We extract 50% overlapping patches and group them into a batch of 360 images and use the tissue region classifier to predict the probability of each tissue class. The probability map for each WSI is saved. After having the probability map for the WSI, prediction map is calculated by taking the class with the highest probability at each patch.

2.4 Quantification

Tumour–stroma ratio (TSR): TSR is calculated at the WSI level according to the following formula:

$$TSR = \frac{s}{s+t} \tag{1}$$

where s in the number of patches classified as stroma and t is the number of patches classified as tumour.

Tumour-infiltrating lymphocytes (TILs): TILs are the lymphocytes found within or around tumour regions. In this study, we quantify TILAb score as introduced in [10]. TILAb score is a prognostic biomarker which quantifies the TIL abundance based on lymphocytes to tumour ratio and their localisation as defined below.

$$f(x) = \begin{cases} \frac{C}{2} \times \frac{\sum_{i=1}^{m}\sum_{j=1}^{n}(p_{ij}^{l})}{\sum_{i=1}^{m}\sum_{j=1}^{n}(p_{ij}^{t})}, & \text{if } \sum_{i=1}^{m}\sum_{j=1}^{n}(p_{ij}^{t}) \geq 0 \\ 1, & \text{otherwise} \end{cases} \tag{2}$$

where C is a co-localisation measure. A WSI is divided into m × n equal-sized grids. We experimented with grid cells of sizes from 4 to 40 in increments of 4 for this quantification.

2.5 *Survival Analysis*

Survival analysis is performed for TSR and TILAb scores to demonstrate their prognostic significance as independent biomarkers for patients OS. Kaplan–Meier [11] and Cox proportional hazards model [12] are used for survival analysis. For TILAb score, to stratify patients into high-risk and low-risk groups, we use the same approach as in [10]; the optimal cut-point is calculated using the modelling subset where the statistical significance of the difference in OS between the two groups is the largest. We also use the same approach to stratify patients into high stroma and low stroma. We did not use the 50% cut-point used by pathologists as TSR in our approach is not calculated by first selecting an area in the WSI where both tumour and stroma are present but is calculated at the WSI level. For univariate Cox model, TSR or TILAb score is used alone; for multivariate Cox model, it is combined with gender, age, tumour stage as covariates. Probability values < 0.05 were considered statistically significant.

3 Results

3.1 *Tissue Region Classification*

Table 2 summarises the accuracy of the five models on the testing set. VGG19 had the highest accuracy of 98.86%. However, we chose Resnet50 for processing the WSIs as we were able to use a larger inference batch size with Resnet50, and this speeds up processing the WSIs. Figure 2 shows the confusion matrix of the Resnet50 model. Figure 3 shows the t-SNE representation of the testing set based on deep layer activations of the CNN [13]. It shows that the CNN was able to learn image features that allow for the separation of nine tissue classes.

Table 2 Performance on classifiers on the testing set

Classifier	Accuracy (%)
Resnet50	97.58
VGG	98.86
DenseNet	96.49
SqueeseNet	96.08
Inception v3	94.0

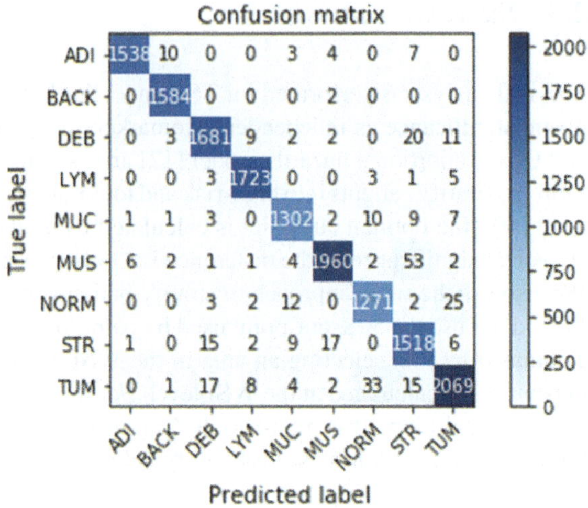

Fig. 2 Confusion matrix

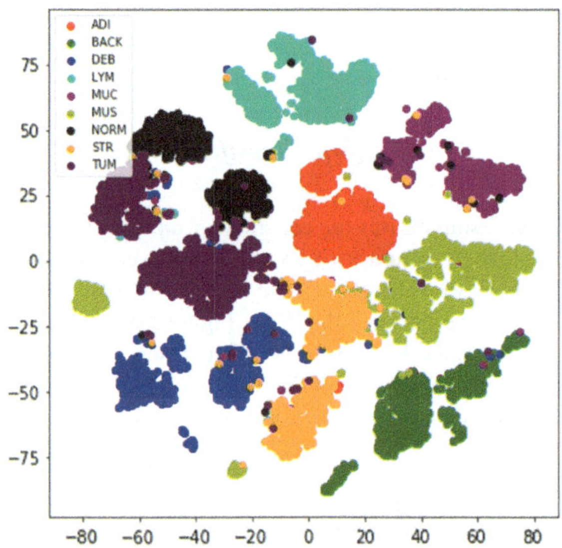

Fig. 3 t-SNE representation

3.2 *Whole Slide Image Analysis*

We applied the tissue classifier to the WSIs from the COAD patients cohort from TCGA. Figure 4 shows an example of a classification map for a WSI. It is interesting to see here how CNNs can decompose the structure of the WSI.

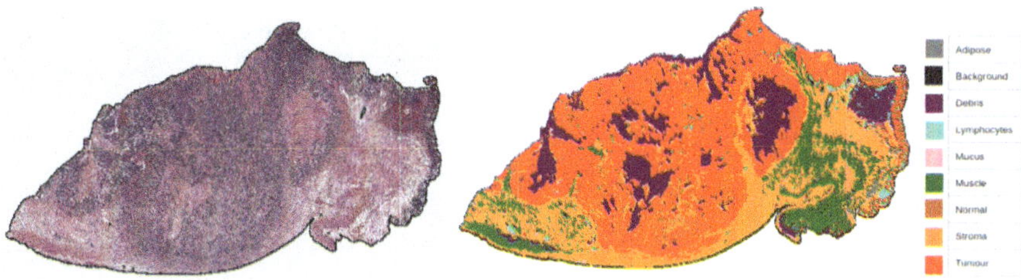

Fig. 4 Classification map

3.3 *Survival Analysis*

Cox proportional hazard models were used to investigate the prognostic significance of TSR and TILAb score for OS survival by conducting univariate and multivariate analysis. We also used Kaplan–Meier (KM) curves to plot the survival functions for OS models for different parameters. We experimented with ten different grid cell sizes for TILAb score. The TILAb score was statistically significant for all sizes, so we report the results of size 20. Table 3 lists the concordance indices for the univariate and multivariate Cox models for OS. The high c-index is an evidence of the predictive ability of the model. The model with the highest predictive power is TSR when combined with age, gender and stage. It has a c-index of 80.5%. Models using TSR achieve higher c-index as compared to models using TILAb score.

The Kaplan–Meier curves in Fig. 5 show the ability of TSR score to stratify patients into stroma-high (poor prognosis) and stroma-low patients. KM curve for the stage of the cancer is also shown in Fig. 5.

Results of univariate analysis of prognostic significance of age, gender, stage, TSR and TILAb score are shown in Fig. 6. Cox proportional hazards model was employed for univariate analysis. Gender does not show any significant correlation with OS. Both TSR and TILAb score are statistically statistical significant with p-values less than 0.05. However, the confidence interval range of hazard ratios is larger for TILAb score. Each of TSR and TILAb score prognostic values was investigated in multivariate Cox models when combined with gender, age and stage as covariates

Table 3 Concordance index for OS models

TSR		TILAb score	
Covariates	C-Index (%)	Covariates	C-Index (%)
TSR	63.64	TILAb score	51.73
TSR, age, gender, stage	80.5	TILAb score, age, gender, stage	78.41
TSR, age	68.9	TILAb score, age	61.66
TSR, gender	63.58	TILAb score, gender	54.02
TSR, stage	77.0	TILAb score, stage	75.25

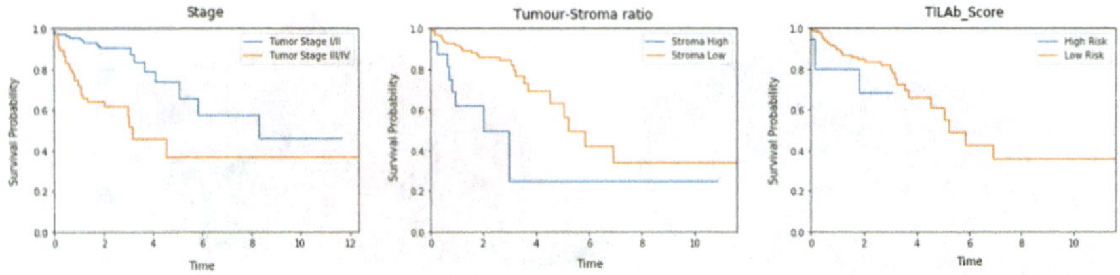

Fig. 5 Kaplan–Meier(KM) curves for overall survival of colon cancer patients on test subset. The optimal cut-point values for TSR and TILAb score are 0.61 and 0.037, respectively

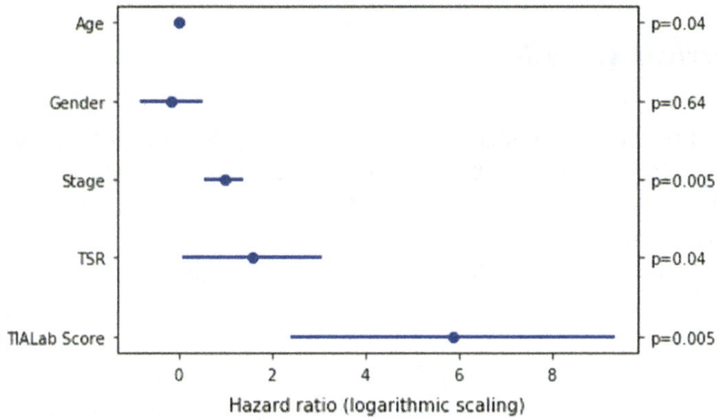

Fig. 6 Univariate analysis of parameters

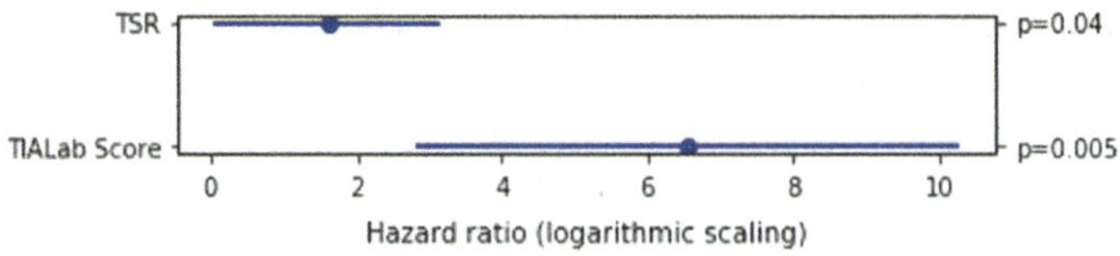

Fig. 7 Multivariate analysis of TSR, TILAb scores, combined with stage, gender, age

as shown in Fig. 7. Results also show that they are both statistically significant, but confidence interval range of hazard ratios is larger for TILAb score, similar to univariate analysis.

4 Discussion and Conclusion

Understanding tumour, stromal and lymphocytes patterns in tissue slides and quantifying them into biomarkers help in developing patients' cancer treatment plan. In this study, we investigated the prognostic value of TSR and TILAb scores on a cohort of TCGA colon cancer patients. After segmenting a WSI into different tissue types in a patch-based approach, localising and counting these tissue types is the way for quantifying the digital biomarkers. Both TSR and TILAb score have shown to be prognostic factors for OS models for the cohort chosen in this study.

We quantified the TSR at the whole slide image level and analysed the prognostic value of this biomarker using univariate and multivariate Cox proportional model. It was shown to be a statistically significant biomarker, especially when combined with the tumour stage, patients' age at diagnosis and gender. We also quantified TILs in this study, and they were also shown to be statistically significant in predicting patients' overall survival. The multivariate Cox model with TSR had a higher concordance index than the multivariate model with TILAb score.

The automated methods for quantifying biomarkers in WSIs have the advantage of being reproducible, while quantification as done by pathologists is subjective with observer variability. Moreover, automated methods help pathologists speed up the process of evaluating tissue slides for routine diagnostics.

References

1. Cancer Research UK. Bowel cancer statistics. https://www.cancerresearchuk.org/health-professional/cancer-statistics/statistics-by-cancer-type/bowel-cancer. Accessed 14 Aug 2019
2. G.W. van Pelt et al., Scoring the tumor-stroma ratio in colon cancer: procedure and recommendations. Virchows Archiv **473**(4), 405–412 (2018)
3. Oscar G.F. Geessink et al., Computer aided quantification of intratumoral stroma yields an independent prognosticator in rectal cancer. Cell. Oncol. **42**(3), 331–341 (2019)
4. J.N. Kather, N. Halama, A. Marx, 100,000 histological images of human colorectal cancer and healthy tissue. Apr. 2018. https://doi.org/10.5281/zenodo.1214456.. https://doi.org/10.5281/zenodo.1214456
5. K. He et al., Deep residual learning for image recognition, in *Proceedings of the IEEE Conference on Computer Vision and Pattern Recognition*, pp. 770–778 (2016)
6. K. Simonyan, A. Zisserman, Very deep convolutional networks for large-scale image recognition. arXiv:1409.1556 (2014)
7. G. Huang et al., Densely connected convolutional networks, in *Proceedings of the IEEE Conference on Computer Vision and Pattern Recognition*, pp. 4700–4708 (2017)
8. F.N. Iandola et al., SqueezeNet: AlexNet-level accuracy with 50x fewer parameters and 0.5 MB model size. arXiv:1602.07360 (2016)
9. C. Szegedy et al., Rethinking the inception architecture for computer vision, in *Proceedings of the IEEE Conference on Computer Vision and Pattern recognition*, 2818–2826 (2016)
10. M. Shaban et al., *Prognostic significance of automated score of tumor infiltrating lymphocytes in oral cancer* (2018)
11. E.L. Kaplan, P. Meier, Nonparametric estimation from incomplete observations. J. Am. Stat. Associat. 53.282, 457–481 (1958)
12. D.R. Cox, Regression models and life-tables. J. Royal Statistical Soc. Ser. B (Methodological) 34.2 187–202 (1972)
13. L.V. Maaten, G. Hinton, Visualizing data using t- SNE. J. Mach. Learn. Res. **9**, pp 2579–2605 (2008)

An Array P System Based on a Variant of Pure 2D Context-Free Grammars

P. S. Azeezunnisha, S. Hemalatha, Sastha Sriram, and Atulya K. Nagar

Abstract The P system model in the area of membrane computing is known to be a convenient framework for handling problems in different areas of application. Several models of array P systems have been developed for dealing with the problem of generation of picture arrays in the area of two-dimensional picture languages. Here, we introduce an array P system which uses in its regions the rewriting rules of pure 2D context-free grammars in the (r/d) derivation mode in order to evolve array objects. We study the generative power of this new P system model by comparing it with certain other picture array generating models.

Keywords Membrane computing · P system · Picture languages · Pure 2D context-free grammar

1 Introduction

The idea of bringing together the two areas of membrane computing [2, 5, 8] and picture array grammars was considered in [1] by developing an array P system for dealing with the problem of handling information structured as two-dimensional

P. S. Azeezunnisha
Department of Mathematics, J.B.A.S. College for Women, Chennai 600018, India
e-mail: hajma29112008@gmail.com

S. Hemalatha
Department of Mathematics, S.D.N.B. Vaishnav College For Women, Chennai 600044, India
e-mail: hemrav2008@gmail.com

S. Sriram (✉)
Department of Mathematics, School of Arts, Science and Humanities, SASTRA Deemed University, Tanjore 613401, India
e-mail: sriram.discrete@gmail.com

A. K. Nagar
School of Mathematics, Computer Science and Engineering, Liverpool Hope University, Liverpool L16 9JD, UK
e-mail: atulya.nagar@hope.ac.uk

© Springer Nature Singapore Pte Ltd. 2020

A. K. Nagar et al. (eds.), *Soft Computing for Problem Solving 2019*,
Advances in Intelligent Systems and Computing 1139,
https://doi.org/10.1007/978-981-15-3287-0_5

(2*D*) objects or arrays [3]. On the other hand, a simple and at the same time expressive 2D grammar model, known as pure 2D context-free grammar ($P2DCFG$), was introduced by Subramanian et al. [9] for generating rectangular picture array languages. Subsequently, a number of studies on this model and its variants [4, 11] have been done, establishing theoretical properties as well as applications to generation of picture patterns. Here, we consider the variant [11] of $P2DCFG$, called $(r/d)P2DCFG$, that involves a (r/d) mode of derivation in the sense of rewriting all the symbols in the rightmost column or in the lowermost row of a rectangular picture array by strings of equal length, thereby maintaining the rectangular shape. We introduce an array P system with objects as rectangular arrays in the regions and groups of context-free rules with (r/d) mode of rewriting as in a $(r/d)P2DCFG$ and investigate the generative power of this new array P system by comparing it with certain other picture array grammars.

2 Preliminaries

For notions on formal language theory, we refer to [6], and for array grammars, we refer to [3].

An alphabet, Σ, is a finite set of symbols. A word or a string $w = w_1 w_2 \dots w_n (n \geq 1)$ over Σ is a finite sequence of symbols w_i, $(1 \leq i \leq n)$ from Σ. The length of a word w is denoted by $|w|$. The set of all words over Σ is denoted by Σ^* which includes the empty word λ. For a given word $w = w_1 w_2 \dots w_n$, we denote by $^t w$ the vertical word $\begin{matrix} w_1 \\ \vdots \\ w_n \end{matrix}$ such that $^t(^t w) = w$. We set $^t\lambda$ as λ itself.

A rectangular $m \times n$ picture array A over an alphabet Σ is of the form

$$A = \begin{matrix} a_{11} & \cdots & a_{1n} \\ \vdots & \ddots & \vdots \\ a_{m1} & \cdots & a_{mn} \end{matrix}$$

where each $a_{ij} \in \Sigma$, $1 \leq i \leq m$, $1 \leq j \leq n$. The set of all rectangular arrays over Σ is denoted by Σ^{**} which includes the empty array λ and $\Sigma^{++} = \Sigma^{**} - \{\lambda\}$. The operations of *column concatenation* and *row concatenation* of arrays in Σ^{**}, denoted, respectively, by $\circ$ and $\diamond$ are defined such that, for any $A, B \in \Sigma^{**}$, $A \circ B$ is defined only when A and B have the same number of rows and $A \diamond B$ is defined only when A and B have the same number of columns.

We now recall the pure 2D context-free grammar introduced in [9].

Definition 1 ([9]) A pure 2D context-free grammar ($P2DCFG$) is a 4-tuple

$$G = (\Sigma, P_1, P_2, \mathcal{M}_0)$$

where

(i) Σ is a finite set of symbols;
(ii) $P_1 = \{c_i \mid 1 \leq i \leq m\}$, where c_i is called a column rule table; each c_i is a finite set of context-free rules of the form $a \rightarrow \alpha, a \in \Sigma, \alpha \in \Sigma^*$ such that for any two rules $a \rightarrow \alpha, b \rightarrow \beta$ in c_i, we have $|\alpha| = |\beta|$, i.e., α and β have equal length;
(iii) $P_2 = \{r_j \mid 1 \leq j \leq n\}$, where r_j is called a row rule table; each r_j is a finite set of rules of the form $c \rightarrow {}^t\gamma, c \in \Sigma, \gamma \in \Sigma^*$ such that for any two rules $c \rightarrow {}^t\gamma, d \rightarrow {}^t\delta$ in r_j, we have $|\gamma| = |\delta|$;
(iv) $\mathcal{M}_0 \subseteq \Sigma^{**} - \{\lambda\}$ is a finite set of axiom arrays.

A picture array $B \in \Sigma^{**}$ is derived from a picture array $A \in \Sigma^{**}$ in G, if B is obtained from A either (*i*) by rewriting in parallel all the symbols in a column of A, each symbol by a rule in some column rule table or (*ii*) rewriting in parallel all the symbols in a row of A, each symbol by a rule in some row rule table. All the rules used to rewrite a column (or row) are to belong to the same table. We denote such a derivation by $A \Rightarrow B$.

The picture language generated by G is the set of picture arrays $L(G) = \{M \in \Sigma^{**} \mid M_0 \Rightarrow^* M$ for some $M_0 \in \mathcal{M}_0\}$. The family of picture languages generated by $P2DCFGs$ is denoted by $P2DCFL$.

We now recall the variant of $P2DCFG$ with (r/d) mode of derivation [11] in which, unlike in a $P2DCFG$, all the symbols in the rightmost column (respectively, the lowermost row) of an array are only rewritten in a step of rewriting, using the rules of a column rule table (respectively, row rule table) with the tables of rules as in a $P2DCFG$.

Definition 2 ([11]) Let $G = (\Sigma, P_1, P_2, \mathcal{M}_0)$ be a $P2DCFG$ with the components as in Definition 1. A (r/d) mode of derivation of a picture array B from A in G is a derivation in G such that only the rightmost column or the lowermost row of A is rewritten using, respectively, the column rule tables or the row rule tables to yield B. We denote such a derivation by $A \Rightarrow_{(r/d)} B$. The generated picture language is defined as in the case of a $P2DCFG$ but with (r/d) mode of derivations. The family of picture languages generated by $P2DCFGs$ under (r/d) mode of derivations is denoted by $(r/d)P2DCFL$. We write $(r/d)P2DCFG$ to refer to $P2DCFG$ with (r/d) mode of derivations.

We illustrate with an example.

Example 1 Consider an $(r/d)P2DCFG$ $G = (\Sigma, P_1, P_2, \{M_0\})$ where $\Sigma = \{x, y, b\}$, $P_1 = \{c\}$, $P_2 = \{r\}$ with $c = \{x \rightarrow bx, y \rightarrow xy\}$, $r = \left\{x \rightarrow \begin{matrix} b \\ x \end{matrix}, y \rightarrow \begin{matrix} x \\ y \end{matrix}\right\}$ and $M_0 = \begin{matrix} b\,x \\ x\,y \end{matrix}$.

G generates a picture language L_{xy} consisting of arrays A of size $(m, n), m \geq 2, n \geq 2$ with $p(m, n) = y$; $p(m, j) = x$, for $1 \leq j \leq n-1$; $p(i, n) = x$, for $1 \leq i \leq m-1$;

$$\begin{array}{cccccc} b & b & b & b & b & x \\ b & b & b & b & b & x \\ b & b & b & b & b & x \\ b & b & b & b & b & x \\ x & x & x & x & x & y \end{array}$$

Fig. 1 A picture array in the language L_{xy}

$$M_0 = \begin{array}{cc} b & x \\ x & y \end{array} \Rightarrow_{(r/d)} \begin{array}{ccc} b & b & x \\ x & x & y \end{array} \Rightarrow_{(r/d)} \begin{array}{ccc} b & b & x \\ b & b & x \\ x & x & y \end{array} \Rightarrow_{(r/d)} \begin{array}{cccc} b & b & b & x \\ b & b & b & x \\ x & x & x & y \end{array}$$

Fig. 2 A sample derivation under (r/d) mode in Example 1

$p(i, j) = b$, otherwise. A member of L_{xy} is shown in Fig. 1. A sample derivation in $(r/d)P2DCFG$ G starting from the axiom array M_0 and using the tables c, r, c in this order is shown in Fig. 2. The application of the column rule table c rewrites all symbols in the rightmost column in parallel, and likewise, the application of the row rule table r rewrites all symbols in the lowermost row. We note that the application of a column table or a row table is independent of each other as in a $P2DCFG$, and so it cannot maintain any proportion between the number of columns and the number of rows in any generated picture array.

Remark 1 In another variant of $P2DCFG$ with (l/u) mode of derivation, introduced in [4] and called $(l/u)P2DCFG$, all the symbols in the leftmost column (respectively, the uppermost row) of an array are only rewritten in a step of rewriting, using the rules of a column rule table (respectively, row rule table) with the tables of rules as in a $P2DCFG$. The corresponding family of picture array languages generated by $(l/u)P2DCFG$ is denoted by $(l/u)P2DCFL$. The $(l/u)P2DCFG$ is a 2D analogue of context-free grammar with leftmost rewriting while $(r/d)P2DCFG$ is a 2D analogue of context-free grammar with rightmost rewriting [11].

3 Array-Rewriting *P* System with Pure Context-Free Rules and (r/d) Mode of Derivation

Motivated by the array-rewriting P system model introduced in [10] with the regions containing rectangular arrays as objects and tables of rules of a pure 2D context-free grammar [9], we define here a variant with the regions containing rectangular arrays as objects and tables of context-free rules as in a $P2DCFG$, but array-rewriting is done in (r/d) mode of derivation as in $(r/d)P2DCFG$.

Definition 3 An array-rewriting P system (of degree $m \geq 1$) with pure 2D context-free rules and (r/d) mode of derivation is a construct

$$\Pi = (V, \mu, F_1, \ldots, F_m, R_1, \ldots, R_m, i_o), \text{ where}$$

V is the alphabet, μ is a membrane structure with m membranes labelled in a one-to-one way with $1, 2, \dots, m$; $F_1, \dots, F_m$ are finite sets of rectangular arrays over V associated with the m regions of μ; $R_1, \dots, R_m$ are finite sets of column tables or row tables of context-free rules over V (as in a $(r/d)P2DCFG$) associated with the m regions of μ with the application of the column rule table which is only on the rightmost column of the picture array rewritten while the application of the row rule table is only on the lower most row; the tables have attached targets *here, out, in, in*$_j$ (in general, *here* is omitted), and finally, i_o is the label of an elementary membrane of μ (the output membrane).

A computation in Π is defined in the same way as in an array-rewriting P system with the successful computations being the halting ones: each rectangular array in each region of the system, which can be rewritten by a column/row table of rules (rewriting as in a $(r/d)P2DCFG$) associated with that region (membrane), should be rewritten; this means that one table of rules is applied; the array obtained by rewriting is placed in the region indicated by the target associated with the table used (*here* means that the array remains in the same region, *out* means that the array exits the current membrane; *in* (respectively, *in*$_j$) means that the array is immediately sent to one of the directly lower membranes (respectively, with label j), nondeterministically chosen if several exist there; if no internal membrane exists, then a table with the target indication *in* cannot be used). A computation is successful only if it stops, that is, a configuration is reached where no table of rules can be applied to the existing arrays. The result of a halting computation consists of rectangular arrays over V placed in the membrane with label i_o in the halting configuration. The set of all arrays computed or generated by such a system Π is denoted by $AL((r/d)\Pi)$. The family of all array languages $AL((r/d)\Pi)$ generated by such systems Π as above, with at most m membranes, is denoted by $AP_m((r/d)P2DCFG)$.

Example 2 Consider the picture language L_{sq} consisting of square-sized arrays A of the language L_{xy} in Example 1, i.e., pictures A of size (n, n), $n \geq 2$ with $A(n, n) = y$; $A(n, j) = x$, for $1 \leq j \leq n-1$; $A(i, n) = x$, for $1 \leq i \leq n-1$; $A(i, j) = b$, otherwise. We construct an array-rewriting P system Π_{sq} with three membranes and in the regions $P2DCFG$ kind of rules in (r/d) mode of derivation. Π_{sq} is given by

$$\Pi = (V, \mu, F_1, F_2, F_3, R_1, R_2, R_3, 3),$$

where

(i) $V = \{x, y, b\}$
(ii) $\mu = [_1 [_2]_2 [_3]_3]_1$
(iii) $F_1 = \{\begin{matrix} b & x \\ x & y \end{matrix}\}$, $F_2 = F_3 = \emptyset$
(iv) R_1 consists of a column rule table c and a row rule table r_1; R_2 consists of a row rule table r_2; $R_3 = \emptyset$; The tables of rules are as follows: $c = \{x \to bx, y \to xy\}(in_2)$, $r_1 = \{x \to \begin{matrix} b \\ x \end{matrix}, y \to \begin{matrix} x \\ y \end{matrix}\}(out)$, $r_2 = \{x \to x, y \to y\}(in_3)$

In the array-rewriting P system Π_{sq}, the skin membrane labelled 1 has inside it, two elementary membranes labelled 2 and 3. Region 3 is the output membrane. Initially, region 1 contains the array $\begin{matrix} b & x \\ x & y \end{matrix}$. If the column rule table c is applied, then a column $\begin{matrix} b \\ x \end{matrix}$ is inserted to the left of the rightmost column and the array $\begin{matrix} b & b & x \\ x & x & y \end{matrix}$ is sent to the region 2. Here, the row rule table r_1 can be applied which inserts a row $b\ b\ x$ just above the last row and sends the array to the region labelled 1. The process can be repeated or the row rule table r_2 can be applied in region 2 which will send the array to the output region 3 without making any change in the array. Since there is no table of rules in region 3, the array remains in the region 3 and is collected in the language generated by Π_{sq}. Thus, the system Π_{sq} generates the language L_{sq}.

We now examine the generative power of the array-rewriting P system with $P2DCFG$ kind of rules in (r/d) mode of derivation.

Theorem 1 $AP_1((r/d)P2DCFG) \subset AP_2((r/d)P2DCFG)$.

Proof The inclusion $AP_1((r/d)P2DCFG) \subseteq AP_2((r/d)P2DCFG)$ follows from the definition of the family $AP_m((r/d)P2DCFG)$.

The proper inclusion can be proved by considering the picture array language L_c consisting of square arrays A of size $n \times n, n \geq 2$, with $A(n, n) = c$, $A(i, n) = a$, for $1 \leq i \leq n - 1$, $A(n, j) = a$, for $1 \leq j \leq n - 1$ and $A(i, j) = b$, otherwise. This language L_c cannot be generated by $(r/d)P2DCFG$ type of rules as the application of the column rule and row rule tables are independent and hence cannot generate only square-sized arrays. Hence, $L_c \notin AP_1((r/d)P2DCFG)$ as the P systems generating languages in $AP_1((r/d)P2DCFG)$ have only one membrane containing all the column or row rule tables. But, L_c can be generated by an array-rewriting P system Π having two membranes with pure 2D context-free rules and (r/d) mode of derivation where $\Pi = (V, \mu, F_1, F_2, R_1, R_2, 2)$, with

(i) $V = \{a, b, e, c\}$
(ii) $\mu = [_1\ [_2\ ]_2]_1$
(iii) $F_1 = \{M_0 = \begin{matrix} b & a \\ a & e \end{matrix}\}$, $F_2 = \emptyset$
(iv) R_1 consists of two column rule tables c_1, c_2; R_2 consists of a row rule table r.

The tables of rules are as follows: $c_1 = \{a \to ba, e \to ae\}(in)$, $c_2 = \{a \to a, e \to c\}(in)$, $r = \left\{a \to \begin{matrix} b \\ a \end{matrix}, e \to \begin{matrix} a \\ e \end{matrix}\right\}(out)$. In a computation in Π, an application of the rules of c_1 will add a column to the left of the rightmost column of M_0 yielding the array $\begin{matrix} b & b & a \\ a & a & e \end{matrix}$, and due to target indication *in* of c_1, the array is then sent to region 2. The rules of the row rule table r can be applied adding a row above the last row of the array, thus yielding the array $\begin{matrix} b & b & a \\ b & b & a \\ a & a & e \end{matrix}$. The process can be repeated. If in region

$$\begin{array}{ccc} b & \cdots & b\ a \\ \vdots & \ddots & \vdots\ \vdots \\ b & \cdots & b\ a \\ a & \cdots & a\ c \end{array}$$

Fig. 3 An array in L_c

1, the rules of the column table c_2 are applied, then the array of the form shown in Fig. 3 is generated and is sent to region 2, where it is collected in the language L_c.

Thus the system Π generates the language L_c.

Theorem 2 *(i)* $AP_2((r/d)P2DCFG) - (r/d)P2DCFL \neq \emptyset$
(ii) $AP_2((r/d)P2DCFG) - (l/u)P2DCFL \neq \emptyset$.

Proof The picture array language L_c considered in the proof of Theorem 1 belongs to the family $AP_2((r/d)P2DCFG)$ but cannot be generated by $(r/d)P2DCFG$ or $(l/u)P2DCFG$ as the application of the rules of a column rule table and the rules of a row rule table can be independently done in both (l/u) mode or (r/d) mode, and hence, L_c cannot be generated as L_c contains only square-sized arrays.

Controlling the application of the tables of rules in a $(r/d)P2DCFG$ by a control language is considered in [11]. The family of picture languages generated by $(r/d)P2DCFG$ with a regular control language is denoted by $(R)(r/d)P2DCFL$. We now consider a non-trivial picture array language L_d [11] consisting of square picture arrays A of size $(n+2) \times (n+2), n \geq 0$, with $A(i, j) = a$ for $i = j$ and $A(i, j) = b$ for $i \neq j$. A member of L_d is shown in Fig. 4.

Theorem 3 *The picture array language L_d belongs to $AP_3((r/d)P2DCFG) \cap (R)(r/d)P2DCFL$.*

Proof The $(R)(r/d)P2DCFG$ given by $G = (\{a, b, e\}, \{c\}, \{r\}, \{\begin{smallmatrix} a & b \\ b & a \end{smallmatrix}\})$ with the column rule table $c = \{a \rightarrow ea, b \rightarrow bb\}$ and the row rule table $r = \{a \rightarrow \begin{smallmatrix} b \\ a \end{smallmatrix}, e \rightarrow \begin{smallmatrix} a \\ b \end{smallmatrix}, b \rightarrow \begin{smallmatrix} b \\ b \end{smallmatrix}\}$ and the control language $\mathcal{C} = \{(cr)^n \mid n \geq 0\}$ generates L_d with the sym-

$$\begin{array}{cccccc} a & b & b & b & b & b \\ b & a & b & b & b & b \\ b & b & a & b & b & b \\ b & b & b & a & b & b \\ b & b & b & b & a & b \\ b & b & b & b & b & a \end{array}$$

Fig. 4 A member of L_d

bol e as the control character. In fact, a derivation starts with the axiom array $\begin{matrix} a & b \\ b & a \end{matrix}$, and the rules of the column rule and row rule tables are applied as specified by a control word of the form $(cr)^n$ yielding a picture array which will belong to L_d. No other array can be generated and hence $L_d \in (R)(r/d)P2DCFL$.

The picture array language L_d also belongs to $AP_3((r/d)P2DCFG)$ being generated by the following array P system Π_d with $P2DCFG$ rules and (r/d) mode of derivation given by $\Pi_d = (V, \mu, F_1, F_2, F_3, R_1, R_2, R_3, 3)$, where $V = \{a, b, e\}$, $\mu = [_1[_2]_2[_3]_3]_1$; Region 3 is the output region. $F_1 = \{\begin{matrix} a & b \\ b & a \end{matrix}\}$; $F_2 = F_3 = \emptyset$. R_1 consists of a column rule table c and a row rule table r_1. R_2 consists of a row rule table r_2. The tables of rules are $c = \{b \to bb, a \to ea\}(in_2)$, $r_1 = \{b \to b, a \to a\}(in_3)$ $r_2 = \{b \to \begin{matrix} b \\ b \end{matrix}, a \to \begin{matrix} b \\ a \end{matrix}, e \to \begin{matrix} a \\ b \end{matrix}(out)\}$.

Starting with the initial array in region 1, if the rules of the column rule table c are applied generating the array $\begin{matrix} a & b & b \\ b & e & a \end{matrix}$, the array is sent to region 2. In this region 2, the row rule table r_2 is applied and the resulting array is sent back to the region 1. The process can be repeated. If the rules of the table r_1 are applied in region 1, the array generated of the form $\begin{matrix} a & b & b \cdots b \\ b & a & b \cdots b \\ \cdots & & \cdots \\ \cdots & & \cdots \\ b & b & b \cdots a \end{matrix}$ is sent to the output region. Thus, the array P system generates L_d.

4 Conclusion

An array P system that evolves rectangular picture arrays in its regions using $P2DCFG$ kind of rules in (r/d) mode of derivation is introduced, and its picture array language generative power is studied. It will be of interest to compare this new array P system model with other picture generating models such as the model in [7].

Acknowledgements The authors thank the reviewers for useful comments.

References

1. R. Ceterchi, M. Mutyam, Gh Păun, K.G. Subramanian, Array-rewriting P systems. Nat. Comput. **2**, 229–249 (2003)
2. M. Gheorghe, Gh Păun, M.J. Perez-Jimenez, G. Rozenberg, Research frontiers of membrane computing: open problems and research topics. Int. J. Found. Comput. Sci. **24**(5), 547–623 (2013)

3. D. Giammarresi, A. Restivo, Two-dimensional languages, in *Handbook of Formal Languages*, vol. 3, ed. by G. Rozenberg, A. Salomaa (Springer, Berlin, 1997), pp. 215–267
4. Z. Krivka, C. Martn-Vide, A. Meduna, K. G. Subramanian, A variant of pure two-dimensional context-free grammars generating picture languages, in *International Workshop on Combinatorial Image Analysis* (2014), pp. 123–133
5. Gh. Păun, *Computing with Membranes: An Introduction* (Springer, Berlin, 2002)
6. G. Rozenberg, A. Salomaa (eds.), *Handbook of Formal Languages*, vol. 1–3 (Springer, Berlin, 1997)
7. G. Samdanielthompson, A.K. Nagar, N.G. David, G. Zhang, K.G. Subramanian, Insertion based picture array P systems, in *Proceedings of the Asian Conference on Membrane Computing*, New Zealand. CDMTCS Research Report Series (2018), pp. 116–125
8. K.G. Subramanian, *P Systems and Picture Languages*. Lecture Notes in Computer Science, vol. 4664 (Springer, Berlin, 2007), pp. 99–109
9. K.G. Subramanian, R.M. Ali, M. Geethalakshmi, A.K. Nagar, Pure 2D picture grammars and languages. Discret. Appl. Math. **157**, 3401–3411 (2009)
10. K.G. Subramanian, L. Pan, S.K. Lee, A.K. Nagar, A P system model with pure context-free rules for picture array generation. Math. Comput. Model. **52**(11–12), 1901–1909 (2010)
11. K.G. Subramanian, M. Geethalakshmi, N.G. David, A.K. Nagar, Picture array generation using pure 2D context-free grammar rules, in *Lecture Notes in Computer Science*, vol. 9448 (2015), pp. 187–201

Predictions of Weekly Slope Movements Using Moving-Average and Neural Network Methods: A Case Study in Chamoli, India

Praveen Kumar, Priyanka, Ankush Pathania, Shubham Agarwal, Naresh Mali, Ravinder Singh, Pratik Chaturvedi, K. V. Uday, and Varun Dutt

Abstract Landslides and associated slope movements are common occurrences in the hilly regions. In particular, Tangni in Uttarakhand state between Pipalkoti and Joshimath has experienced a number of landslides in the recent past. Prior research has used a certain moving average and machine-learning (ML) algorithms to predict slope movements. However, a comparison of these methods for real-world slope movements has been less explored. The primary objective of this paper was to compare a seasonal autoregressive integrated moving average (SARIMA) model, a multilayer perceptron (MLP) model, and a long short-term memory (LSTM) model to predict slope movements recorded at the Tangni landslide in Chamoli, India. Time series data about slope movements from five sensors placed on the Tangni landslide hill were collected daily over a 78-week period from July 2012 to July 2014. Different model parameters were calibrated to the training data (first 62 weeks) and then made to predict the test data (the last 16 weeks). Results revealed that the moving-average models (SARIMA) performed better compared to the ML models (MLP and LSTM) during both training and test. We highlight the implication of using moving-average models for predicting slope movements.

P. Kumar (✉) · Priyanka · A. Pathania · S. Agarwal · V. Dutt
Applied Cognitive Science Lab, Indian Institute of Technology Mandi, Mandi, Himachal Pradesh, India
e-mail: bluecodeindia@gmail.com

Priyanka
e-mail: priyankasihag8993@gmail.com

P. Chaturvedi · K. V. Uday
Defence Terrain Research Laboratory, Defence Research and Development Organization (DRDO), New Delhi, India

N. Mali
Geohazard Studies Laboratory, Indian Institute of Technology Mandi, Mandi, Himachal Pradesh, India

R. Singh
National Disaster Management Authority (NDMA), New Delhi, India

© Springer Nature Singapore Pte Ltd. 2020

A. K. Nagar et al. (eds.), *Soft Computing for Problem Solving 2019*,
Advances in Intelligent Systems and Computing 1139,
https://doi.org/10.1007/978-981-15-3287-0_6

Keywords Landslides · SARIMA · MLP · LSTM · Slope movements · Tangni landslide

1 Introduction

Landslides afflict the Himalayan region, and landslide occurrence is common in hilly areas of India in states of Himachal Pradesh and Uttarakhand [1]. These landslides cause a huge damage to life and property [2]. Thus, we need to monitor, predict, and alert people about slope movements on hills prone to landslides [3]. A way of predicting slope movements is via statistical or machine-learning (ML) algorithms [4]. Such statistical or ML algorithms take certain data attributes as input and predict the value of other attributes of interest [5].

Several statistical and ML algorithms have been proposed in the literature [6–16]. Two of the most popular algorithms are either based upon moving average (statistical) or based upon neural networks (ML) [9–13]. For example, the seasonal autoregressive integrated moving average (SARIMA) algorithm is a popular moving-average-based method, where the past values in a time series are used to predict future values [15, 16]. Similarly, algorithms like MLP and LSTM are neural network-based algorithms.

Prior research involving slope movement prediction has used both the neural network-based methods [6–8] and moving-average-based methods [9–16]. For example, Refs. [15, 16] have used some moving-average algorithms to predict slope movements. Similarly, Refs. [17, 18] used the MLP and LSTM models to predict the slope movements.

Although several ML algorithms have been used in the literature to predict slope movements in the recent past [6–16], a comparison of moving-average-based methods and neural network-based methods has been less discussed. The primary objective of this paper is to compare certain moving-average-based methods (SARIMA) with neural network-based methods (MLP and LSTM) in their ability to predict slope movements in an active landslide site in the Himalayan mountains. Specifically, we use a 2-year time series data about weekly slope movements from the Tangni landslide in Chamoli, India, for our study.

In what follows, first, we detail the background literature on the use of different moving-average and ML algorithms for predicting slope movements. Next, we detail the working of the SARIMA, MLP, and LSTM algorithms and the method of training these algorithms to data from the Tangni landslide. Finally, we present our results from different algorithms and discuss the implication of using statistical and ML methods for slope movement predictions.

2 Background

Several past research studies have used neural network (ML) models [6–8] for predicting slope movements as well as finding different triggering factors for such movements. For example, Refs. [6, 7] proposed a novel neural network technique called ensemble of extreme learning machine (E-ELM) and investigated the interactions of different inducing factors affecting slope movements. Next, Ref. [6] improved the ELM model by proposing a MEEMDELM model. The results show that the MEEMDELM model was consistently better than the basic artificial neural networks (ANNs) and the unmodified EEMDELM model in terms of the same measurements. Furthermore, Ref. [8] used a multiple-ANNs switched prediction method on three typical landslides in Three Gorges Reservoir, namely Baishuihe landslide, Bazimen landslide, and Shiliushubao landslide. These authors found that the proposed switched prediction method could significantly improve model generalization compared to the best individual ANN predictor.

Some researchers have also used tree-based models for predicting slope movements [14]. For example, Ref. [14] presented a methodology for the prediction of landslide movements using random forests, a machine-learning algorithm based on regression trees. The random forest method was established based on a time series consisting of 2 years of data on landslide movement, groundwater level, and precipitation gathered from the Kostanjek landslide monitoring system and nearby meteorological stations in Zagreb (Croatia). The validation results showed the capability of the random forest model to predict the evolution of daily displacements for the period up to 30 days.

Reference [19] have compared LSTM models with multilayer perceptron (MLP) and support vector machines for prediction of slope movement in the Baijiabao landslide. Data were divided into trend and periodic components using the empirical mode decomposition. When using one-step-ahead prediction, LSTMs perform better compared to the MLPs and support vector machines. Reference [20] predicted landslides on the China–Nepal Highway. These authors compared four different machine-learning algorithms, namely decision tree (DT), support vector machines (SVMs), backpropagation neural network (BPNN), and long short-term memory (LSTM) models. The LSTMs outperformed the other three models due to their capability to learn time series with long temporal dependencies. Reference [21] developed a MLP to predict the slope movement and applied this MLP model to two landslides in Switzerland (La Chenaula) and France (Salldes). Neural networks are easier to implement and more general than statistical regressive models [21]. In fact, Ref. [21] found the good results with the MLP model.

Furthermore, some researchers have found the moving-average models to perform correctly in predicting slope movements [15, 16]. For example, Ref. [15] used the autoregressive integrated moving average (ARIMA) model, and this model was employed to forecast the accumulative displacement of the Bazimen landslide. Results indicated that the ARIMA method improved the result of traditional static data. In addition, Ref. [16] compared the ARIMA model with the considerable autoregressive (CAR) model and found both these methods to yield good results.

Given that both ARIMA (statistical) and MLP and LSTM (ML) models performed well in accounting for slope movements, in this paper, we formally compare these approaches on a single dataset.

3 Study Area

The study was done on the Tangni landslide in Chamoli district of Uttarakhand, India. The study area covers an area of 0.72 km^2. The study area is located on the northern region of Himalayan at latitudes 30 27 54.3 N and longitudes 79 27 26.3 E, at an altitude of 1450 m (Fig. 1a, b). As seen in Fig. 1b, the landslide is located on National Highway 58, which connects Ghaziabad in Uttar Pradesh near New Delhi with Badrinath and Mana Pass in Uttarakhand. The geology of this area contains the slate and dolomite rocks [3]. The landslide slope is 30 above the road level and 42 below the road level. The nearby area is filled with oak and pinewood trees. There have been several previous landslides in this area causing roadblocks and economic losses to tourism [17].

Data were collected from the Tangni landslide at a daily scale between July 1, 2012, and July 1, 2014, across five different boreholes. These five boreholes are represented by different colors in Fig. 1b (red borehole 1, green borehole 2, yellow borehole 3, blue borehole 4, and pink borehole 5). Each borehole contained five sensors at different depths (3, 6, 9, 12, and 15 m). Data from some of these five boreholes were used for evaluating different moving average and neural network-based methods.

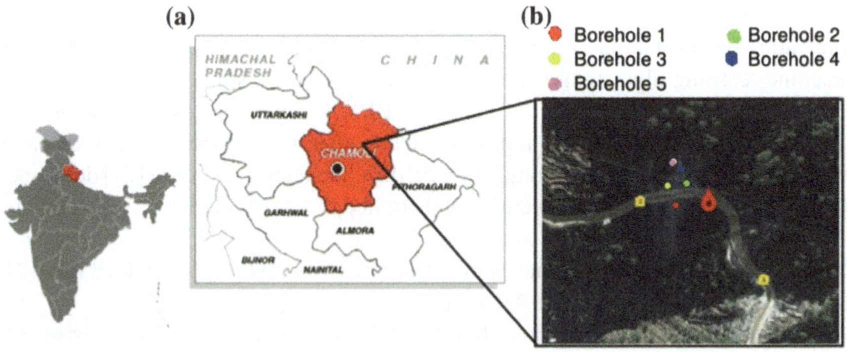

Fig. 1 Tangni landslide in Uttarakhand. *Source* Google maps

4 Methodology

4.1 Data Preprocessing

Data from Tangni landslide in Chamoli, India, were obtained from the Defence Terrain Research Laboratory, Defence Research and Development Organization. The monitoring system deployed in each borehole possessed inclinometer sensors at different depths (3, 6, 9, 12, and 15 m). These sensors measured tilt in mm per m units (essentially the angle the inclinometer tilts). Each inclinometer sensor was a 0.5 m long sensor rod that was installed vertically in a borehole. The monitoring system at Tangni landslide had five sensors per borehole across five boreholes. Thus, in total, there are 25 sensors across 5 boreholes. Figure 2 shows a set of inclinometer sensors installed in a casing at different depths in one of the boreholes at Tangni. As shown in Fig. 2, if there is a tilting movement (θ) of the inclinometer of length L sinθ, then the horizontal displacement in the tilting direction is. For better understanding, we converted the displacement in mm per m units into a θ angle (in degrees), where 1 mm/m displacement equaled 0.05729°.

Data of tilt angle in degree were available from the initial period when sensors were installed. We chose one sensor from each borehole that possessed the largest average tilt angle over a two-year period. That is because the sensor with the largest average tilt angle was the one closest to the slip surface. Thus, the data were reduced to five time series, where each time series represented the relative tilt per borehole from the sensor that moved the most in the borehole across the two-year period. As the daily data possessed sparse movement, we averaged the daily tilt over a weekly period to yield 78 weeks of average tilt data per time series. Figure 3a–e represents the average relative tilt per week from five sensors across five boreholes (one sensor

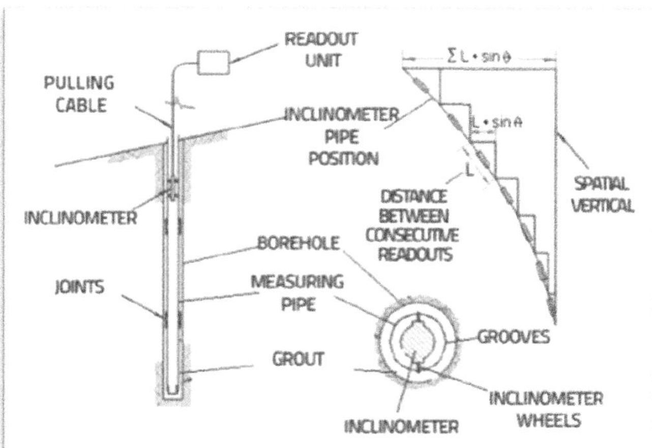

Fig. 2 Inclinometer sensors installed in a borehole

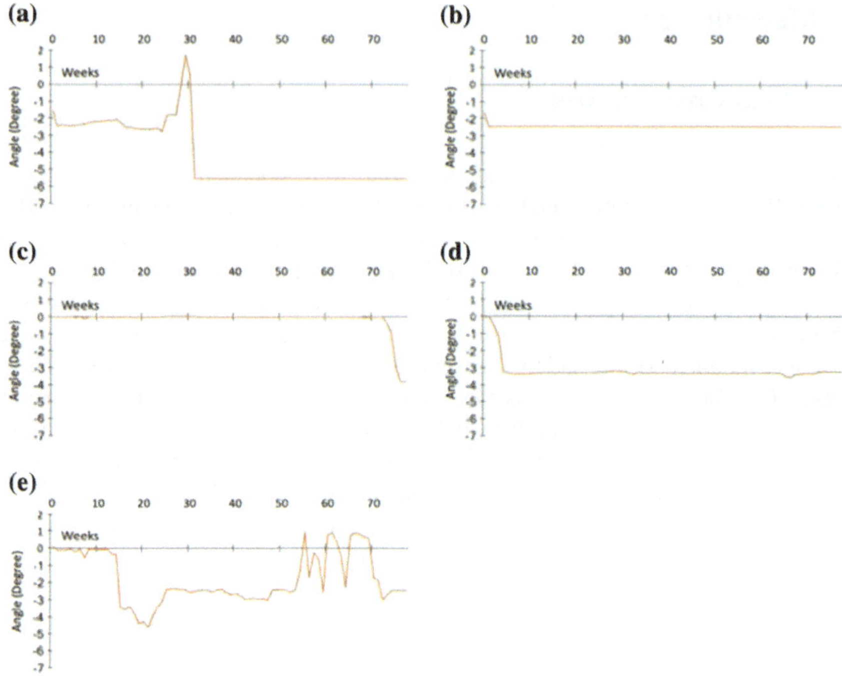

Fig. 3 Average tilt angle in degrees across five sensors (one per borehole). **a** Borehole 1 and 3 m sensor. **b** Borehole 2 and 12 m sensor. **c** Borehole 3 and 6 m sensor. **d** Borehole 4 and 15 m sensor. **e** Borehole 5 and 15 m sensor

per borehole). These sensors caused the maximum average tilt across 78 weeks. These five time series were used to compare different statistical and ML models.

By convention, a negative tilt angle was a downhill motion and a positive tilt angle was an uphill motion of the sensor. As seen in Fig. 3a, a downhill motion starts from −2 in the 29th week movement suddenly becomes larger (2) in positive afterword in 30th week movement becomes −6 and remains constant until last week. The data were split in 80:20 ratios (sixty-two weeks for training and the last sixteen weeks for testing) across different statistical and ML algorithms.

4.2 Seasonal Autoregressive Integrated Moving Average

Seasonal autoregressive integrated moving average (SARIMA) is a statistical predicting method popular for univariate time series data that may contain trend and seasonal components. It predicts the time series by describing the autocorrelations in data [18].

Stationarity of time series: A time series with constant values over time for mean, variance, and autocorrelation is stationary. Most statistical forecasting methods assume that a time series can be made approximately stationary using mathematical transformations such as differencing [22]. The first step of building a SARIMA model is stationarizing the data.

Autoregressive (AR): An autoregressive component in the SARIMA model predicts a variable at present state by passing the previous values of the same variable to the model. Thus, an autoregressive model is defined as follows:

$$y_t = c + \phi_1 y_{t-1} + \phi_2 y_{t-2} + \cdots + \phi_p y_{t-p} + \varepsilon_t \tag{1}$$

where p is the autoregressive trend parameter, ε_t is white noise, and $y_{t-1}, y_{t-2} \ldots y_{t-p}$ denote the movement data of the past weeks [18].

Moving average (MA): A moving-average component in the SARIMA model uses previous prediction errors in a regression model, which is given in Eq. (2). A moving-average model is defined as follows:

$$y_t = c + \theta_1 \varepsilon_{t-1} + \theta_2 \varepsilon_{t-2} + \cdots + \theta_q \varepsilon_{t-q} + \varepsilon_t \tag{2}$$

where q is the moving-average trend parameter, ε_t is white noise, and $\varepsilon_{t-1}, \varepsilon_{t-2} \ldots \varepsilon_{t-q}$ are the error terms at past weeks.

If we combine auto-regression (AR) in Eq. (1) and a moving average (MA) in Eq. (2) model on stationary data, we obtain a non-seasonal ARIMA model, which is defined as follows:

$$y_t = c + \phi_1 y'_{t-1} + \cdots + \phi_p y'_{t-p} + \theta_1 \varepsilon_{t-1} + \cdots + \theta_q \varepsilon_t \tag{3}$$

The SARIMA model builds upon the ARIMA model by integrating seasonal parts to the ARIMA model. The seasonal parts of an ARIMA model can have an AR factor, a MA factor, and an order of difference term. All these factors in seasonal data operate across multiples of the number of lagged periods in a season. In SARIMA, the three trend elements that require calibration are the trend AR order p, the trend difference order d, and trend MA order q. Additional four seasonal elements that require calibration are the seasonal AR order P, the seasonal difference order D, the seasonal MA order Q, and the number of time steps m for a single seasonal period. A SARIMA model performs differencing of order D at a lag equal to the number of seasons m to remove additive seasonal effects. As with lag 1 differencing to remove the trend, the lag m differencing introduces a moving-average term.

4.3 *Multilayer Perceptron (MLP)*

A multilayer perceptron (MLP) is feed-forward artificial neural network. At a minimum, it has three layers: an input layer, a hidden layer, and an output layer. Each node

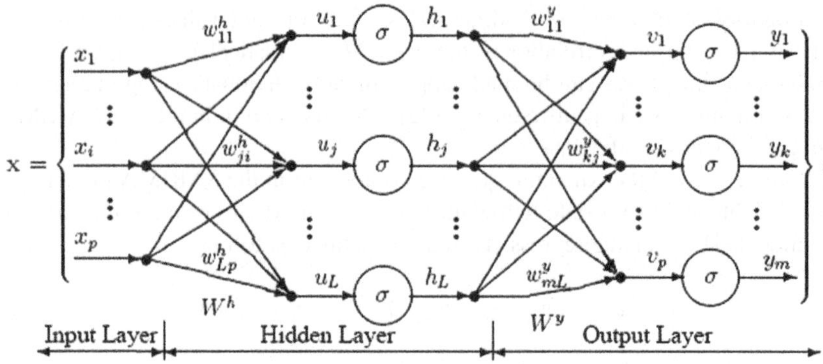

Fig. 4 Architecture graph of MLP [26]

of one layer is connected to all other nodes of the next layer, except the input nodes. Every layer uses a activation function (see Fig. 4). It applies a supervised learning technique called backpropagation for training [23]. The connection between every two nodes of MLP has a certain weight. The weights between all the nodes are updated based on the error (the difference between the target value and predicted value), which is done through backpropagation.

At output node j, error in nth training point is represented by

$$e_j(n) = d_j(n) - y_j(n) \tag{4}$$

where d is the actual value and y is the predicted value. The node weights are changed such that the error in the entire output is defined as

$$\varepsilon(n) = \frac{1}{2}\sum_j e_j^2(n) \tag{5}$$

Using the gradient descent method, the change in each weight is

$$\Delta\, w_{ji}(n) = -\eta \frac{\delta\varepsilon(n)}{\delta v^j(n)} y_i(n) \tag{6}$$

where y_i is the output of the previous neuron and 'η' is the learning rate of MLP, which is selected such that the weights converge without any oscillations. The derivative $\frac{\delta\varepsilon(n)}{\delta v^j(n)}$ at output layer is a function of the error term (Eq. 5) and derivative of the activation function.

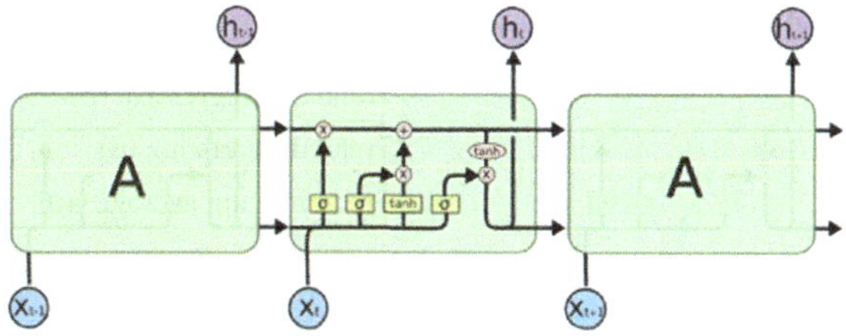

Fig. 5 Repeating module in an LSTM containing four interacting layers [27]

4.4 *Long Short-Term Memory (LSTM)*

LSTMs belong to the family of recurrent neural networks (RNNs) that are capable of learning long-term dependencies [24]. RNNs are a chain of units, containing loops which act as a memory element used to learn the temporal behavior of the data. In standard RNNs, each unit has a very simple structure, such as a single tanh layer. But RNNs often fail to learn the long-term dependencies and have a vanishing gradient problem.

To reduce these problems, LSTMs are preferred and these units give significant improvements over RNNs. LSTMs have a cell state and three gates which provide them with the power to selectively learn, unlearn, or retain information from each of the units [25] (see Fig. 5).

The cell state in LSTM acts as a conveyor belt which helps the information to flow through the units with only a few linear interactions, without being altered [4]. Each unit has an input, output, and a forget gate which can add or remove the information to the cell state. The forget gate decides which information from the previous cell state should be forgotten for which it uses a sigmoid function. The input gate controls the information flow to the current cell state using a point-wise multiplication operation of sigmoid and tanh, respectively. Finally, the output gate decides which information should be passed on to the next hidden state. The sigmoid function outputs numbers between zero and one, describing how much of each component should be let through (see Fig. 5) [25].

4.5 *Optimization of Model Parameters*

4.5.1 SARIMA

This model has eight different parameters, namely p, d, q, P, D, Q, m, and Trend. Here, p means the number of autoregressive (AR) terms, d means the order of differencing (I), and q means the size of the moving-average (MA) window. Table 1 shows the range of variations for different parameters in the SARIMA model. The trend parameter has four different values, where absent means no trend, constant means constant (horizontal) trend, linear means linear trend, and finally, the constant with

Table 1 Parameters for SARIMA

Parameters	Range of values
Trend autoregressive (p)	[0, 1, 2]
Trend differencing (d)	[0, 1]
Trend moving average (q)	[0, 1, 2]
Trend	[Absent, Constant, Linear, Constant with Linear Trend]
Seasonal autoregressive (P)	[0, 1, 2]
Seasonal differencing (D)	[0, 1]
Seasonal moving average (Q)	[0, 1, 2]
Seasonal periods (m)	[0, 1]

linear trend means there is both a constant and linear trend. The *m* parameter is the number of time steps for a single seasonal period. A parameter value of zero means that the parameter is absent in the model. A reason for using the SARIMA model was that it allows one to account for a seasonal trend present in the time series.

4.5.2 MLP

This model has seven hyper-parameters (see Table 2). Here, the number of inputs is the number of lag observations to use as input to the model, number of nodes is the number of nodes to use in hidden layer, number of epochs is the number of time to expose the model to the whole training dataset, and number of batches is the number of samples within an epoch after which the weights are updated. We put the parameters' value in MLP as per the following: for number of inputs = 3, for number of nodes = 50, for number of epochs = 100, and for number of batch = 50.

Table 2 Parameter optimization of MLP and LSTM

Parameters	Range of values
Number of hidden layers	3
Number of inputs	1, 3, 6
Number of nodes	50, 100, 150, 200
Number of epochs	50, 100, 150
Batch size	1, 25, 50, 60
Activation functions	Rectified Linear (ReLu), Linear, Sigmoid
Optimizer	Adam, Root Mean Square Prop (RMSProp)

4.5.3 LSTM

LSTMs have a similar architecture to MLP, except that all the hidden layers are not dense. The first hidden layer was a LSTM layer. This model has four layers, one input layer, two hidden layers, and one output layer. The various ranges of values of parameters taken by this model are presented in Table 2. We put the parameters' values in LSTM as per the following table.

5 Results

Each model was calibrated to each time series, independently. Table 3 shows the root-mean-squared error (RMSE) results of training of LSTM, MLP, and SARIMA models on the training data across the five boreholes. As can we see in Table 3, among all the algorithms, the LSTM and SARIMA models performed the best and second best for training data. As can we see in Table, among all the algorithms, the LSTM and SARIMA models performed the best and second best for training data.

Tables 4, 5, and 6 show the optimized values of different parameters of in the SARIMA, MLP, and LSTM models, respectively. For example, different SARIMA models showed nonzero seasonal Q parameter and non-seasonal q parameter. In certain cases, the seasonal P, the non-seasonal p, and the non-seasonal d parameters possessed nonzero values.

Table 7 shows the RMSEs from different models across different boreholes in the last 16 weeks of test data. As can be seen in the table, the SARIMA algorithm

Table 3 RMSE of different algorithms in the training dataset

Algorithm	Root-mean-squared error (RMSE) in degree of angle					
	Borehole 1	Borehole 2	Borehole 3	Borehole 4	Borehole 5	Average
	3 m	12 m	6 m	15 m	15 m	
LSTM	0.93	0.00	0.00	0.01	0.90	0.37
SARIMA	0.86	0.00	0.03	1.04	1.11	0.71
MLP	1.03	0.00	0.03	3.00	0.90	0.99

Table 4 Optimized parameters for SARIMA

Sensor location		Best set of parameters
Borehole	Depth (m)	[(p, d, q), (P, D, Q, m), Trend]
1	3	[(0, 1, 0), (0, 0, 1, 0), 'Absent']
2	12	[(0, 0, 1), (1, 0, 1, 1), 'Constant']
3	6	[(0, 1, 1), (0, 0, 0, 0), 'Absent']
4	15	[(2, 0, 1), (0, 0, 1, 0), 'Absent']
5	15	[(1, 0, 0), (0, 0, 0, 0), 'Absent']

Table 5 Optimized parameters for MLP

Parameter	Range of values				
	Borehole 1 3 m	Borehole 2 12 m	Borehole 3 6 m	Borehole 4 15 m	Borehole 5 15 m
Number of hidden layers	1	1	1	1	1
Number of inputs	3	3	3	1	6
Number of nodes	75	100	100	100	150
Number of epochs	100	100	100	100	100
Batch size	50	25	1	25	50
Activation functions	ReLu	ReLu	ReLu	ReLu	ReLu
Optimizer	Adam	Adam	Adam	Adam	Adam

Table 6 Optimized parameters for LSTM

Parameter	Range of values				
	Borehole 1 3 m	Borehole 2 12 m	Borehole 3 6 m	Borehole 4 15 m	Borehole 5 15 m
Number of hidden layers	12	2	2	2	2
Number of inputs	3	1	1	1	1
Number of nodes	150	50	50	150	100
Number of epochs	100	100	100	100	100
Batch size	1	25	1	25	50
Activation functions	ReLu	ReLu	ReLu	ReLu	ReLu
Optimizer	Adam	Adam	Adam	Adam	Adam

Table 7 RMSE of different algorithms in the test dataset

Algorithm	Root-mean-squared error (RMSE) in degree of angle					
	Borehole 1 3 m	Borehole 2 12 m	Borehole 3 6 m	Borehole 4 15 m	Borehole 5 15 m	Average
SARIMA	0.00	0.00	0.52	0.06	1.06	0.33
LSTM	0.00	0.00	0.73	0.07	1.05	0.37
MLP	0.00	0.00	0.76	0.06	1.05	0.38

performed best compared to the MLP and LSTM algorithms, and MLP is the second best algorithm.

Figure 6 shows the fits of the SARIMA algorithm to the time series data across the five boreholes in the training and test datasets. Overall, these results are reasonably good with very small RMSE values.

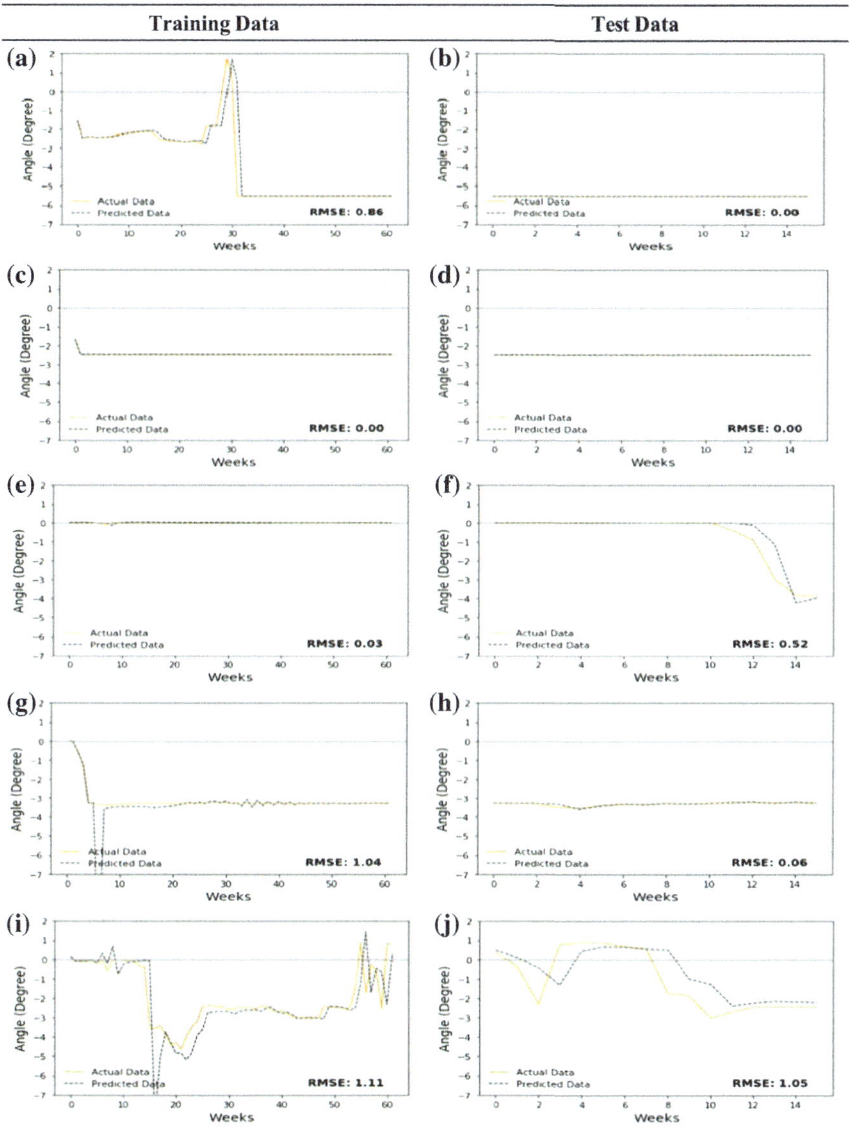

Fig. 6 Relative angle (in degree) over training data (62 weeks) and test data (16 weeks) from the best performing SARIMA algorithm. **a** and **b**: borehole 1, 3 m depth. **c** and **d**: borehole 2, 12 m depth. **e** and **f**: borehole 3, 6 m depth. **g** and **h**: borehole 4, 15 m depth. **i** and **j**: borehole 5, 15 m depth

6 Discussion and Conclusions

The main focus of machine-learning algorithms in landslide mitigation is to predict the movements of the slope so that the lives can be saved. In this work, we applied moving-average-based models (SARIMA) and neural network-based models (MLP and LSTM) on weekly slope movement data from the Tangni 1 and slides in Chamoli, India. We used training test data split, root-mean square error (RMSE), and one-step-ahead walk forward validation for model comparison. All models were calibrated on the first 80 could generate the slope movements in the following week given the history of movements in previous weeks. Our results revealed that the moving-average-based models (SARIMA) do not perform better during training time in comparison with neural network-based model (MLP and LSTM), but performed well during testing data. One likely reason for our results is the seasonal effect in the movement data. Due to the seasonal effect in the data, the movement was high during the rainy season. Furthermore, the movement was low during the dry season periods. In future, we will use the LSTM, MLP, and SARIMA models with multivariate data for predicting the slope movement.

In this paper, we were able to show that moving-average-based methods outperform neural network-based methods for real-world slope movement predictions.

Acknowledgements The project was supported by grants (awards: IITM/NDMA/VD/184, IITM/DRDO-DTRL/VD/179 and IITM/DCoN/VD/204) to Varun Dutt. We are also grateful to the Indian Institute of Technology Mandi for providing computational resources for this project.

References

1. R.K. Pande, Landslide problems in Uttaranchal, India: issues and challenges. Disaster Prev. Manag.: Int. J. 15(2), 247–255 (2006)
2. S. Parkash, Historical records of socio-economically significant landslides in India. J. South Asia Disaster Stud. (2011)
3. P. Chaturvedi, S. Srivastava, P.B. Kaur, Landslide early warning system development using statistical analysis of sensors? Data at Tangni Landslide, Uttarakhand, India, in *Proceedings of Sixth International Conference on Soft Computing for Problem Solving* (Springer, Singapore, 2017), pp. 259–270
4. O. Korup, A. Stolle, Landslide prediction from machine learning. Geol. Today **30** (2014). https://doi.org/10.1111/gto.12034
5. R.O. Duda, P.E. Hart, D.G. Stork, *Pattern Classification* (Wiley, New York, 2012), p. 2014
6. C. Lian, Z. Zeng, W. Yao, H. Tang, Ensemble of extreme learning machine for landslide displacement prediction based on time series analysis. Neural Comput. Appl. **24**(1), 99–107 (2014)
7. Y. Cao, K. Yin, D.E. Alexander, C. Zhou, Using an extreme learning machine to predict the displacement of step-like landslides in relation to controlling factors. Landslides **13**(4), 725–736 (2016)
8. C. Lian, Z. Zeng, W. Yao, H. Tang, Multiple neural networks switched prediction for landslide displacement. Eng. Geol. **186**, 91–99

9. C. Zhou, K. Yin, Y. Cao, B. Ahmed, Application of time series analysis and PSO? SVM model in predicting the Bazimen landslide in the Three Gorges Reservoir, China. Eng. Geol. **204**, 108–120 (2016)
10. Z. Liu, J. Shao, W. Xu, H. Chen, C. Shi, Comparison on landslide nonlinear displacement analysis and prediction with computational intelligence approaches. Landslides **11**(5), 889–896 (2014)
11. A. Brenning, Spatial prediction models for landslide hazards: review, comparison and evaluation. Nat. Hazards Earth Syst. Sci. **5**(6), 853–862 (2005)
12. X. Zhu, Q. Xu, M. Tang, W. Nie, S. Ma, Z. Xu, Comparison of two optimized machine learning models for predicting displacement of rainfall-induced landslide: a case study in Sichuan Province, China. Eng. Geol. **218**, 213–222 (2017)
13. C.H. Zhu, G.D. Hu, Time series prediction of landslide displacement using SVM model: application to Baishuihe landslide in Three Gorges reservoir area, China. in *Applied Mechanics and Materials*, vol. 239. (Trans Tech Publications, 2013), pp. 1413–1420
14. M. Krka, D. Poljari, S. Bernat, S.M. Arbanas, Method for prediction of landslide movements based on random forests. Landslides **14**(3), 947–960 (2017)
15. G.H. Duan, R.Q. Niu, A method of dynamic data mining for landslide monitoring data. J. Yangtze River Sci. Res. Inst. **30**(5), 10 (2013)
16. L.I. Qiang, L.I. Duan-you, Research of dynamic predication technique for landslide displacement monitoring. J. Yangtze River Sci. Res. Inst. 6 (2005)
17. India News, Landslides near Badrinath in Uttarakhand, 13 August 2013, https://www.indiatvnews.com/news/india/landslides-near-badrinath-inuttarakhand-26296.html. Accessed 7 Apr 2019
18. D. Asteriou, S.G. Hall, ARIMA models and the Box? Jenkins methodology. Appl. Econom. **2**(2), 265–286 (2011)
19. S. Xu, R. Niu, Displacement prediction of Baijiabao landslide based on empirical mode decomposition and long short-term memory neural network in Three Gorges area, China. Comput. Geosci. **111**, 87–96 (2018)
20. L. Xiao, Y. Zhang, G. Peng, Landslide susceptibility assessment using integrated deep learning algorithm along the china-nepal highway. Sensors **18**(12), 4436 (2018)
21. F.D. Bortoloti, T.W. Rauber, Comparison of computational intelligence techniques in the optimization of a neural network topology for prediction of landslides in Vitria-Es (Brazil)
22. R.J. Hyndman, G. Athanasopoulos, Forecasting: principles and practice. OTexts (2018)
23. D.E. Rumelhart, G.E. Hinton, R.J. Williams, Learning internal representations by error propagation (1986)
24. L.R. Medsker, L.C. Jain, Recurrent neural networks: design and applications (1999)
25. S. Hochreiter, J. Schmidhuber, Long short term memory (1997)
26. A. Faghfouri, M.B. Frish, Robust discrimination of human foot-steps using seismic signals (2011)
27. C. Olah, Understanding LSTM network (2015)

[illegible]

Two-Stage History Matching for Hydrology Models via Machine Learning

Dewi Tjia, Ritu Gupta, and Muhammad Alam

Abstract The reliability of a hydrological model (HydroM) highly depends on how well the model parameters are estimated through the history matching (HisM) process. Direct HisM (DHisM) that calibrates input parameters by iteratively executing a HydroM is widely applied in water resource estimation. The computational time of DHisM is prohibitive, as a single run of the HydroM may take several hours. In practice, calibration accuracy is compromised to arrive at a solution. Therefore, it is desirable to develop a proxy model that can replace HydroM in the HisM process. In this study, we propose a two-stage HisM, wherein we first develop a proxy model for HydroM using artificial neural network techniques. Next we apply ant colony optimisation (ACO_R) and robust parameter estimation (ROPE) methods for calibrating the parameter of HydroM. This methodology is illustrated for the Dandalups catchment of Western Australia to calibrate the five global parameters of Land Use Change Incorporated Catchment (LUCICAT) by matching 33 annual daily streamflow peaks. The results reveal that replacing the LUCICAT by proxy model reduces the computational time by more than 90% with similar accuracy to DHisM and shows higher consistency (via standard deviation of RMSE) and reduction of parameter uncertainty.

Keywords History matching · Parameter estimation · Ant colony optimisation · Robust parameter estimation method · Proxy model · Machine learning · Artificial neural network

1 Introduction

Hydrological model (HydroM) is an important tool for water resource estimation and management. This model comprises mathematical equations that describe complex

D. Tjia · R. Gupta (✉)
Department of Mathematics and Statistics, Curtin University, Bentley, WA 6102, Australia
e-mail: R.Gupta@curtin.edu.au

M. Alam
Department of Water, Perth, WA, Australia

© Springer Nature Singapore Pte Ltd. 2020
A. K. Nagar et al. (eds.), *Soft Computing for Problem Solving 2019*,
Advances in Intelligent Systems and Computing 1139,
https://doi.org/10.1007/978-981-15-3287-0_7

natural physical process of a catchment. Inputs of HydroM are climatic data and catchment characteristic. Catchment characteristics need to be calibrated to match historical response, such as streamflow, salinity and groundwater level. The process of adjusting input parameters of a catchment to reduce the discrepancy between model prediction and historical data is known as history matching (HisM).

HisM is a challenging task as HydroM is often nonlinear and computationally complex. For the past four decades, many researchers have conducted studies to automate HisM for challenging problems [6, 23]. HisM methods typically work on an initial set of input parameters, and this set is recursively updated to minimise an objective function describing differences between the observed data and model response. In this process, a HydroM is run several hundred and thousand times to obtain an optimal set of input parameters. According to Tjia's study [20], for a HisM approach, such as ant colony algorithm and robust parameter estimation, the computational time for HisM of a HydroM takes a few hours to a few days. For more complex HydroM, for instance, when physical processes are represented by grid cells, a HydroM becomes computationally expensive [13]. Hence, the implementation of HisM using a HydroM directly, which is named as direct HisM (DHisM) approach, could be computationally challenging. We propose two-stage HisM by first developing a proxy model for a HydroM, then using this proxy model and HisM methods to find a set of optimal input parameter of HydroM.

The idea of replacing computer model like LUCICAT, by an accurate, robust and computationally convenient proxy model, can revolutionise the study of computer codes for any aspect of predictive analytics. The machine learning methods, like artificial neural network (ANN), can be successfully used to develop good proxy model. In this paper, we demonstrate that the implementation of this approach offers the significant improvement in the efficiency of HisM process, with no compromise on the final results.

A very well-known modelling method, namely ANN based on resilient backpropagation with one, two and three layers, is used to develop proxy models in this paper. ANN has been widely applied in many fields such as science, engineering [21, 22] and medical science [24].

We apply two HisM methods, namely ant colony optimisation for continuous domain problem (ACO_R) and robust parameter estimation (ROPE). Both of these methods have been successful in solving the HisM problems in various fields of computer modelling. ACO_R which is inspired by foraging behaviour of real ants has been used for HisM in water resource management [1, 19] and engineering application [10, 16]. ROPE, based on geometrical position of a data in the data set, has been used in rain runoff modelling [12, 17].

This paper is organised as follows: Section 2 describes materials and methods used in this paper including the study area, the hydrology model (LUCICAT), description of ANN model and HisM methods (ACO_R and ROPE) and simulation settings. Section 3 provides results and discussions. Finally, conclusions and recommendations for further research are presented in Sect. 4.

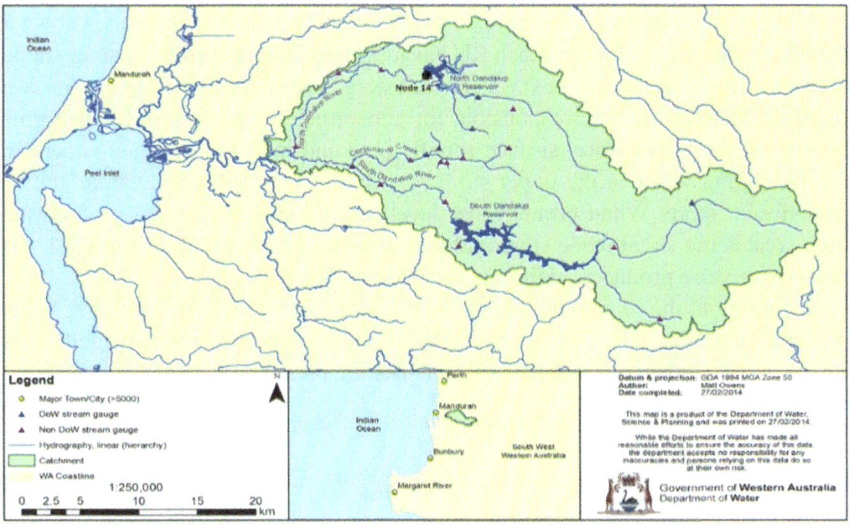

Fig. 1 Dandalups [15]

2 Materials and Methods

2.1 Study Area

Land Use Change Incorporated Catchment (LUCICAT) HydroM is developed for the Dandalups catchment, Western Australia. The Dandalups is an important area as the two reservoirs within the catchment, namely North Dandalup and South Dandalup, supply Perth's annual water consumption. The location of the Dandalups, surface water bodies and gauging stations are depicted in Fig. 1. The area of the catchment is about 700 km^2. The catchment has elevation ranging from 10 AHD (Australian Height Datum) to 500 AHD, with an average slope between 1.2 and 16.2%. The catchment has a Mediterranean climate characterised with a hot and dry summer season. The average annual precipitation is about 1,100 mm. The Dandalups has historical rainfall and streamflow data from 1 January 1960 to 14 November 1992 (33 years). HisM, in this study, is based on Node 14 (Fig. 1) since this node is the most influential node for the catchment.

2.2 LUCICAT HydroM

The LUCICAT HydroM [3] is a distributed conceptual HydroM used for predicting streamflow and salinity due to land and climate changes. The model has been used for many applications of catchment modelling in Western Australia [11]. In the LUCICAT model, a catchment is divided into several subcatchments, known as response units (RUs) (Fig. 2). Each RU has five interconnected stores, namely upper store (dry stores and wet stores), subsurface store, streamzone store and groundwater store. These stores are responsible for generating streamflow and transporting the salt. In the upper store, surface runoff (Q_{r1}) and interflow (Q_i) are generated. The remaining water in the upper store percolates to the subsurface store and the groundwater store. When groundwater level reaches stream bed, the groundwater store creates the streamzone store and generates baseflow to the stream (Q_b). The streamzone store produces additional surface runoff (Q_{r2}).

According to the sensitivity analysis test performed by the Department of Water, Western Australia, LUCICAT has five global sensitive parameters that significantly influence the model response [5]. Table 1 lists the parameters, their role and preset uncertainty ranges.

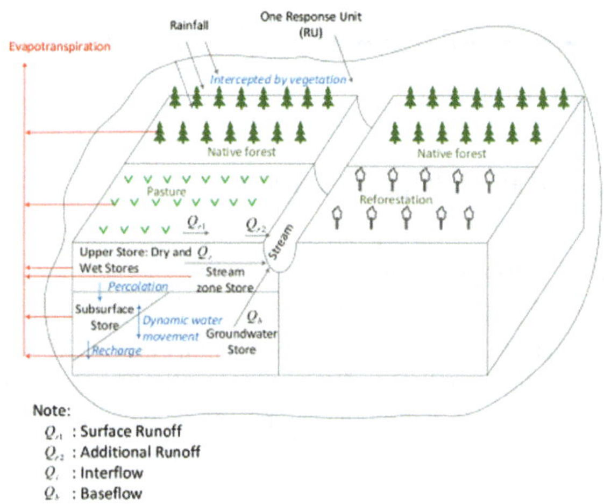

Fig. 2 LUCICAT building model

Table 1 Five global input parameters of LUCICAT model

Parameter	Role	Units	Range
ia	Interflow generation	–	1.8–3.1
K_{uv}	Vertical conductivity of top soil	Mm/d	15.3–27.2
b	Dry water soil moisture exponent	–	0.156–0.56
K_{ll}	Baseflow generation	Mm/d	400–1000
c	Wet water soil moisture exponent	–	0.156–0.56

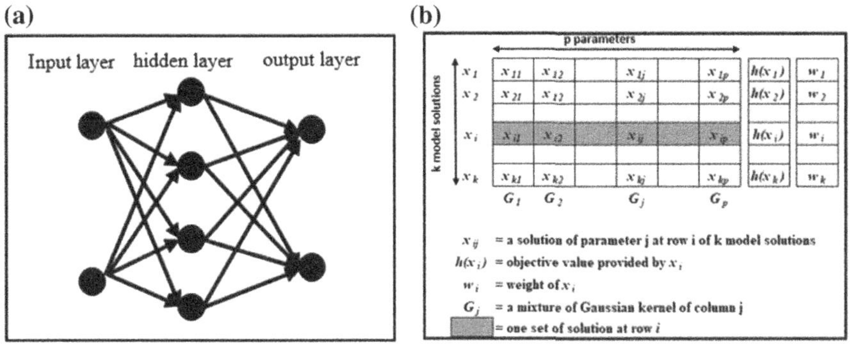

Fig. 3 Example of ANN (**a**) and an ACO_R solution archive (**b**)

2.3 *Artificial Neural Network (ANN)*

ANN method is inspired by the successful parallel working neurons in the human brain. In an ANN, artificial neurons are structured into several layers, which are input, hidden and output layers. Neurons in each layer are connected with synapses. Figure 3a depicts an example of an ANN with one hidden layer. An ANN model is represented by the following relationship:

$$o(\boldsymbol{x}) = g\left(w_0 + \boldsymbol{w}^T \boldsymbol{x}\right) \tag{1}$$

and $g(.)$ is the activation function with characteristic as a bounded non-decreasing and differentiable function. The synaptic weights are iteratively updated to minimise the discrepancy between the predicted output and the actual training response using learning algorithm, such as backpropagation and resilient backpropagation (Rprop). We use Rprop as a learning algorithm to reduce potentially slower learning of weights that lie in the layer far from the output layer [14]. Rprop has been used in energy modelling problem [4] and tourism forecasting [7].

2.4 *History Matching Methods*

Ant Colony Optimisation based on Continuous Domain $(\mathbf{ACO})_\mathbf{R}$

ACO_R was firstly introduced by Socha and Dorigo [18] to handle continuous optimisation problems. The idea of ACO_R lies on the construction of a solution archive, **X** with dimension of k rows (number of solutions) by p columns (number of input parameters). This solution archive is depicted in Fig. 3b. Every cell of the archive has jth unknown input parameter of ith model solution denoted as x_{ij}. For each solution x_i, the objective function $h(x_i)$ weight w_i is computed. The weights are calculated according to the following formula:

$$w_i = \frac{1}{qk\sqrt{2\pi}}\exp\left(\frac{-(i-1)^2}{2q^2k^2}\right) \quad (2)$$

where q is the parameter control for selecting model in the solution archive. A mixture of Gaussian kernels is used to construct new members based on the information of the solution archive. For input parameter *j*, $G_j(\boldsymbol{x})$ comprises weighted sum of k one-dimensional Gaussian functions ($g_{ij}(x)$) expressed as follows:

$$G_j(\boldsymbol{x}) = \sum_{i=1}^{k} w_i \frac{1}{\sigma_{ij}\sqrt{2\pi}} \exp\left(\frac{\left(x-\mu_{ij}\right)^2}{2\left(\sigma_{ij}\right)^2}\right) \quad (3)$$

where $j \in \{1, \ldots, p\}$. $G_j(\boldsymbol{x})$, has parameters a mean vector $\boldsymbol{\mu}_j$ and a vector of standard deviations $\boldsymbol{\sigma}_j$. Each element of $\boldsymbol{\mu}_j$ is defined by the corresponding value in the solution archive, $\mu_{ij} = x_{ij}$ for $i = 1, \ldots, k$. Each component of $\boldsymbol{\sigma}_j$ is calculated by:

$$\sigma_{ij} = \xi \sum_{e=1}^{k} \frac{\left|x_{ej} - x_{ij}\right|}{k-1} \quad (4)$$

where ξ is pheromone evaporation that controls how fast the algorithm forgets the worse solutions. $\mathbf{X}_{\text{new}}$ with matrix size of $m \times p$ is constructed by taking samples from mixture of Gaussian kernels. $\mathbf{X}_{\text{new}}$ is combined with $\mathbf{X}$ and ranked. The m worst solutions are removed from the table to keep the size of archive the same and assign these solutions as $\mathbf{X}$.

Robust Parameter Estimation (ROPE)

ROPE was firstly proposed by Bardossy and Singh [2]. ROPE is a geometrical approach based on data depth. Data depth is a measure of the degree of centrality of a point with respect to a multivariate data set. Many variants of data depth, however, in this study we focus on the halfspace depth. The halfspace depth of points of a point $z \in R^p$ with respect to a p-dimensional data set is defined as the minimum number of data points lying on a closed halfspace plan with boundary of the plane through z.

$$D(z, X) = \text{minimum}\ \#\left(i : u^T x_i \geq u^T z\right) \quad (5)$$

where $X = \{\boldsymbol{x}_i = \left(x_{i1}, \ldots, x_{ip}\right), i = 1, \ldots, n\}$. $\boldsymbol{u}$ ranges over all vector in R^p, $||\boldsymbol{u}|| = 1$.

ROPE initially generates a set of n input parameters and calculates the corresponding objective function. p percent of data with the best objective function values is identified as a set of good performing model parameters. Then, the algorithm searches a set of robust parameters within a good performing model parameter set by using the concept of halfspace depth, to generate m new input parameters.

2.5 Simulation Settings

Historical Response: These are 33 annual daily streamflow peaks for the period between 1960 and 1992. These peaks were considered the most influential in the choice of model parameters.

Proxy Model: ANN Rprop [8] with one, two and three hidden layers were trained using data set of size of 50, 100, and 200 generated using Latin Hypercube Design (LHD). The average performance of ANN Rprop for each size was evaluated by using a testing data set, a size of 100, and the best average performing ANN model was selected for HisM.

History matching: LHD was used to design initial parameter settings for HisM algorithms. LHD with S-optimality criterion of size of 50 was used for ACO_R, while we used LHD with maximin criterion of size of 1000 for ROPE. LHD with S-optimality criterion is computationally extensive when the size of design is large as used for ROPE. LHDs were generated via DoW R package [9]. ACO_R and ROPE were executed to minimise the objective function defined as:

$$\text{RMSE} = \sqrt{\frac{\sum_{j=1}^{33}\left(Y_{\text{flow}_j}^{h} - Y_{\text{flow}_j}^{s}\right)^2}{33}} \tag{6}$$

where $Y_{\text{flow}_j}^{h}$ and $Y_{\text{flow}_j}^{S}$ are the historical streamflow and the proxy model response for jth, $j \in \{1, \ldots, 33\}$. HisM process was repeated until a stopping criterion was achieved. For ACO_R, stopping criteria was a difference of less than 0.005 between the best RMSE of four consecutive iterations. For ROPE, HisM process was terminated if the mean RMSE of two consecutive iterations was less than 0.005. The tuning parameters of HisM methods were adopted from Tjia's study [20]. For ACO_R, $k =$ m $= 50$, $\xi = 0.5$ and $q = 0.05$ and ROPE, the initial settings (n) and new generation of solutions (m) were set each 1000, and the new solutions were selected from the best 10% (p) solutions of previous iterations. The performances of HisM using both algorithms were evaluated in terms of accuracy (mean RMSE), consistency (standard deviation of RMSE), computational time and reduction of initial parameter uncertainty for the six experiments.

2.6 Simulation System

All coding routines of ACO_R and ROPE were written in R statistical software. For DHisM, the LUCICAT model was called and executed in R environment, wherein the algorithms updated the input parameter and calculated RMSE. The simulation was performed using a PC with an Intel® Core™ i5-2500 CPU at 3.3 GHz, 8 Gb of RAM.

3 Result and Discussion

3.1 Performance of Proxy Models

The performances of ANN models for one, two and three hidden layers from various training data sizes are illustrated in Table 2.

Table 2 shows that all ANN models of training data size of 50 have similar average performances (p-value = 0.311). By contrast, when the training data size increases (100 and 200), there are significant differences in mean RMSE across number of layers (p-value = 0.001 and p-value = 0.042). Tukey's multiple comparison demonstrates that ANN from the training data sizes of 100 and 200 has comparable performances for two and three hidden layers and outperform the ANN model with one hidden layer. It is also observed that there is similar consistency in standard deviation of RMSE for all number of hidden layers from training data sizes of 50 and 100. Unlike these results, the ANN (based on the training data size of 200) with three hidden layers is more consistent in RMSE than the models with one or two hidden layers. Comparing ANN performances from different size of data training within the same number of hidden layer, it is clearly seen the performances of ANN improves when the size of data used for model training increases. However, the computational time increases (up to 1000 fold) when the size of training data increases, and ANN

Table 2 Best performances of ANN model from three training data size

Training data	Number of hidden layers			p-value
	One	Two	Three	
Size of 50	h1 = 70	h1 = 10, h2 = 80	h1 = 10, 2 = 50, h3 = 70	
Mean RMSE	0.207	0.199	0.197	0.311
Stdev	0.019	0.015	0.015	0.531
Comp.time (s)	24.0	50.7	89.0	0.000
Size of 100	h1 = 50	h1 = 10, h2 = 50	h1 = 10, 2 = 50, h3 = 70	
Mean RMSE	0.144	0.117_a	0.116_a	0.001
Stdev	0.000	0.007	0.006	0.064
Comp.time (s)	59.5	115.8	148.7	0.000
Size of 200	h1 = 60	h1 = 10, h2 = 50	h1 = 10, 2 = 10, h3 = 100	
Mean RMSE	0.101	0.079_a	0.078_a	0.042
Stdev	0.004_a	0.005_a	0.001	0.031
Comp.time (s)	388.9	435.7	1,224.7	0.000

Note h1, h2, h3 are number of neurons in the first, second and third hidden layers respectively. Models with the same subscript represent pairs with equal means RMSE at 5% level of significance based on Tukey's multiple comparison, while variance is based on F-test for pairwise comparison with Bonferroni correction

Table 3 Performances of ACO_R and ROPE using direct and two-stage approaches

ACO_R with	LUCICAT	ANN with n hidden layers		
		n = 1	n = 2	n = 3
Mean RMSE	3.665	3.671	3.670	3.658
Stdev	0.012	0.017	0.008	0.002
p-value t-test		0.471	0.426	0.273
p-value F-test		0.534	0.371	0.001
ROPE with	LUCICAT	ANN with n hidden layers		
		n = 1	n = 2	n = 3
Mean RMSE	3.675	3.698	3.672	3.671
Stdev	0.012	0.028	0.002	0.003
p-value t-test		0.139	0.529	0.389
p-value F-test		0.058	0.002	0.002

structures are more complex (one to three hidden layers). The gain in accuracy is not consistent with the increase in computational time.

3.2 *History Matching Results*

In order to keep the computational time pragmatic, we only present HisM results using ANN model developed by a training data size of 50. Our results show that HisM performances using ANN model trained by the data size of 50 (mean RMSE and standard deviation RMSE) are similar to those provided by ANN models of the training data sizes of 100 and 200. The comparison of HisM results between DHisM using the LUCICAT model and two-stage HisM approach described above are presented in Table 3. These results were obtained by running six experiments, for which each experiment used 400 model runs. According to t-test, the performances of both HisM methods, namely DHisM and two-stage HisM approach (the ANN with one, two and three hidden layers), are comparable (all p-values t-test > 0.05) for average RMSE. In terms of consistency (standard deviation of RMSE), F-test analysis shows that ACO_R with the ANN three hidden layers is significantly more consistent than ACO_R in DHisM approach (p-value F-test = 0.001). For ROPE, higher consistency in RMSE is achieved when the algorithm coupled with the ANN two (p-value F-test = 0.002) and three hidden layers (p-value F-test = 0.002).

Further results in Table 4 show that two-stage HisM approach significantly reduces the average computational time for DHisM by more than 90% for both ACO_R and ROPE. This concludes that two-stage HisM approach provides a very high efficiency in time for HisM. This procedure may become more advantageous when the HydroM is more complex (using each grid to represent physical hydrological process for instance).

Table 4 Average computational time (s) using LUCICAT and ANN models

	LUCICAT	ANN with n hidden layers		
		n = 1	n = 2	n = 3
ACO_R				
Data acquisition		2,484	2,484	2,484
Model development		24.0	115.8	148.7
HisM process	33,001	0.5	0.6	1.2
Total	33,001	2,508.5	2,600.4	2,633.9
% reduction		92.4	92.1	92.0
ROPE				
Data acquisition		2,484	2,484	2,484
Model development		24.0	115.8	148.7
HisM process	295,074	94.2	146.6	162.2
Total	295,074	2,602.2	2,746.4	2,794.9
% reduction		99.1	99.1	99.1

Another important criterion in evaluating the performances of models and methods is the reduction of initial input parameter uncertainty. Figure 4 illustrates 95% CI for input parameter uncertainty for DHisM and two-stage HisM approach (using ANN model developed from 50 training data sizes). For most of the input parameters, two-stage HisM approach shows significant reduction in the initial range of uncertainty (Table 1), and final values are within the 95% CI of results of direct HisM. However, both algorithms with current settings are observed to have difficulties in reducing input parameter uncertainties produced by the ANN model with one hidden layer.

The calibrated values of the input parameters based on two-stage HisM methods results in a range of streamflow between 22.5 and 22.9 million m^3/year in comparison to the estimate of 22.6 and 22.9 million m^3/year for DHisM, requiring 90% less computational time. The estimates obtained are of similar order.

4 Conclusion

This paper discussed the implementation of two-stage HisM approach using proxy models. The purpose of HisM is to estimate the five sensitive input parameters of the HydroM for the Dandalups catchment by matching the historical 33 annual daily peaks of streamflow. ANN based on Rprop algorithm was used for developing proxy model for the LUCICAT model as input in the process of HisM. HisM was then performed using ACO_R and ROPE HisM methods.

The performances of ANN with one, two and three hidden layers improved on average in terms of RMSE and consistency (standard deviation) as training data size increased. The ANN from training data size of 50 had comparable performances

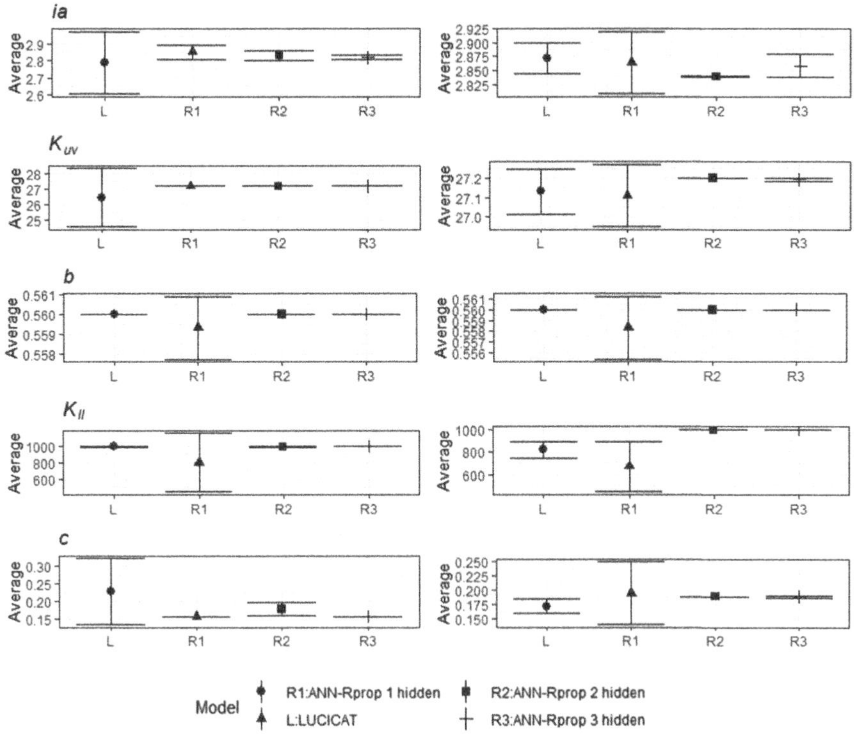

Fig. 4 Reduction of input parameter uncertainty between DHisM and two-stage HisM approach with ACO_R (left) and ROPE (right)

across number of hidden layers. These findings were different from the results of larger training data size (100 and 200), where the ANN with two and three hidden layers performed better than one hidden layer. For ANN, there was a tendency of having similar variation in RMSE across number of hidden layers, when the models came from smaller training data size. In spite of a better performance, ANN models trained by larger training data sets required longer computational time.

Two-stage HisM results had comparable performances with DHisM methods in terms of mean RMSE. In terms of consistency, lower variation in RMSE values occurred for proxy models based on the ANN with two and three hidden layers in comparison to the DHisM. Furthermore, the ranges provided by two-stage HisM approach were much lower than those provided by DHisM, with one exception of two-stage HisM approach using ANN with one hidden layer. Additionally, the application of two-stage HisM approach reduces computational time by more than 90% compared to the DHisM. We had similar experience in implementing two-stage HisM for other catchments. This study shows the successful implementation of HisM using the proxy model for the LUCICAT.

References

1. K. Ananda Babu, R.K. Shrivastav, Developing objective function by ant colony optimization for continuous domains (ACO_R): sondur reservoir. Int. J. Latest Trends Eng. Technol. (IJLTET) **2**, 58–64 (2013)
2. A. Bardossy, S.K. Singh, Robust estimation of HydroM parameters. Hydrol. Earth Syst. Sci. **12**, 1273–1283 (2008)
3. M.A. Bari, A distributed conceptual model for stream salinity generation processes: a systematic data based approach. Doctor of Philosophy thesis, The University of Western Australia (2005)
4. Y. Cui, Z. Yi, J. Duan, D. Shi, Z. Wang, A Rprop-Neural-Network-Based PV Maximum power point algorithm with short-circuit current limitation, in *Electrical Engineering and System Science*. IEEE ISGT NA (2019)
5. Department of Water: LUCICAT user manual (2012)
6. Q. Duan, S. Sorooshian, V. Gupta, Effective and efficient global optimization for conceptual rainfall-runoff models. Water Resour. Res. **28**(4), 1015–1031 (1992)
7. P.O. Fernandez, J.P. Teixeira, J. Ferreira, S. Azevedo, Training neural network by resilient backpropagation algorithm for tourism forecasting. Manag Intell. Syst. 41–46 (2013)
8. S. Fritsch, F. Guenther, M.N. Wright, M. Suling, S.M. Mueller, Package 'neuralnet' version 1.44.2 (2019)
9. U. Groemping, DoE.wrapper: wrapper package for design of experiments functionality, r package version 0.31. https://cran.r-project.org/web/packages/DoE.base/index.html. Accessed 1 Nov 2017
10. Y. Hajizadeh, M. Christie, V. Demyanov, Ant colony optimization for history matching and uncertainty quantification of reservoir models. J. Petrol. Sci. Eng. **77**, 78–92 (2011)
11. L.R. Joyce, The hydrological impacts of climate change and variability in the Murray Hotham catchment, Western Australia. Honours Dissertation, School of Environment Systems Engineering, The University of Western Australia (2007)
12. T. Krausse, J. Cullmann, Toward a more representative parameterization of hydrologic models via synthesizing the strength of particle swarm optimization and robust parameter estimation. Hydrol. Earth Sci. **16**, 603–629 (2012)
13. H. Kuntsmann, A. Heckl, A. Rimmer, Physically based distributed HydroMling of the upper Jordan catchment and investigation of effective model equation. Adv. Geosci. **9**, 123–130 (2006)
14. M. Miller, Rprop-description and implementation details. Technical Report. Institut fur Logic, Komplexitat und Dedutionsysteme, University of Karlsruhe (1994)
15. M. Owen, The Dandalups Catchment scale 1:250,000. Department of Water, Western Australia
16. F. Simsek, F. Mergen, Optimization of induction motors using continuous ant colony systems. Int. J. Recent Trends Eng. Technol. **4**(3), 62–65 (2010)
17. S.K. Singh, A. Bardossy, HydroM calibration by sequential replacement of weak parameter sets using depth function. Hydrology **2**(2), 69–92 (2015)
18. K. Socha, M. Dorigo, Ant colony optimization for continuous domains. Eur. J. Oper. Res. **185**, 1155–1173 (2008)
19. D. Tjia, R. Gupta, M. Alam, Ants do history matching in hydrology, in *Ozwater 2015*, Adelaide, South Australia, 12–14 May 2015
20. D. Tjia, Statistical methods for history matching of HydroM. Doctor of Philosophy thesis, Curtin University, Perth, Western Australia (2016)
21. D. Tumac, Artificial neural network application to predict the sawability performance of large diameter circular saws. Measurement **80**, 12–20 (2016)
22. A.G.R. Vaz, B. Elsinga, W.G.J.H.M. van Sark, M.C. Brito, An artificial neural network to assess the impact of neighbouring photovoltaic systems in power forecasting in utrect, The Netherlands. Renew Energy **85**, 631–641 (2016)

23. Q.J. Wang, The genetic algorithm and its application to calibrating conceptual rainfall-runoff models. Water Resour. Res. **27**(9), 2467–2471 (1991)
24. Y. Wang, J. Li, J. Gu, Z. Zhou, Z. Wang, Artificial neural network for infectious diarrhea prediction using meteorological factors sin Shanghai (China). Appl. Soft Comput. **35**, 280–290 (2015)

An Intelligent System for Diagnosis of Diabetic Retinopathy

Saroj Kr. Biswas, Rohit Upadhya, Nipan Das, Dolly Das, Manomita Chakraborty, and Biswajit Purkayastha

Abstract Diabetic retinopathy (DR) or diabetic eye disease is a medical condition in which damage occurs to the retina due to diabetes mellitus causing blindness. Early detection of DR helps in preventing the onset of blindness in diabetic patients. Numerous intelligent models have been proposed for early detection of DR. However, models incorporating support vector machine (SVM) for early detection of DR have been very rare. Therefore, this paper proposes a model called intelligent system for diabetic retinopathy (ISDR) for early detection of DR using SVM. The fundus images captured with the help of a digital fundus camera (DFC) are used as inputs to the proposed model. The fundus images are initially enhanced and then segmented to extract two most prominent features for the early detection of DR, namely foveal avascular zone (FAZ) and microaneurysms (MA). The extracted features are used as inputs to SVM for classification of the fundus images to classify them as 0—no DR, 1—mild DR, 2—moderate DR, 3—non-proliferative DR (NPDR) and 4—proliferative DR (PDR). The model is validated by the Kaggle dataset. It is observed from the experimental results that early detection of DR is feasible using ISDR.

S. Kr. Biswas (✉) · R. Upadhya · N. Das · D. Das · M. Chakraborty · B. Purkayastha
Department of Computer Science and Engineering, National Institute of Technology Silchar, Silchar 788010, Assam, India
e-mail: bissarojkum@yahoo.com

R. Upadhya
e-mail: rohitupadhya18@gmail.com

N. Das
e-mail: nipandas83@gmail.com

D. Das
e-mail: das.dolly0019@gmail.com

M. Chakraborty
e-mail: mou.look@gmail.com

B. Purkayastha
e-mail: biswajitpurkayastha1960@gmail.com

© Springer Nature Singapore Pte Ltd. 2020

A. K. Nagar et al. (eds.), *Soft Computing for Problem Solving 2019*,
Advances in Intelligent Systems and Computing 1139,
https://doi.org/10.1007/978-981-15-3287-0_8

Keywords Diabetic retinopathy · Retinal lesions · Retinal fundus · Fundus camera · Blindness

1 Introduction

Diabetic retinopathy or diabetic eye disease is a medical condition which damages the retina as a result of diabetes mellitus thus causing blindness. According to statistics, 80% of the people who have had prolonged diabetes for more than 15–20 years suffer from diabetic retinopathy (DR) [1]. It is widely prevalent among the working-age people and causes severe blindness if not detected at an early stage. The motivation behind this work is to propose an intelligent system for the detection of DR for its earlier detection so that DR can be prevented. DR is diagnosed with a comprehensive dilated eye exam. In this test, an ophthalmologist looks for symptoms of the disease in the patient's eyes through the process of intravenous injection in which contrasting agents are poured into the eyes to dilate the pupil and looks through a high-resolution camera such as DFC, which captures images of the rear view of the eye [2]. The ophthalmologist uses these images to determine the severity of the disease. Certain retinal lesions and rupturing of the retinal blood vessels in these images indicate the severity of the disease. However, the manual analysis of retinal images is time-consuming and also requires an expert ophthalmologist. Thus, the need of proper analysis of the retinal images using advanced image processing and enhancement techniques stimulates the importance of an intelligent and automated system. Such a system can take the retinal images as an input and perform image analysis by extracting the important features to determine the earlier symptoms and/or stages of DR. Therefore, this paper proposes a model called intelligent system for diabetic retinopathy (ISDR) for early detection of DR using SVM. The fundus images captured with the help of a digital fundus camera (DFC) are used as inputs to the proposed model. The fundus images are initially enhanced by contrast limited adaptive histogram equalization (CLAHE) and then segmented by Otsu's thresholding and Canny's edge detection to extract two most prominent features, namely foveal avascular zone (FAZ) and microaneurysms (MA). The extracted features are used as inputs to SVM for classification of the fundus images into five classes: 0—no DR, 1—mild DR, 2—moderate DR, 3—non-proliferative DR (NPDR) and 4—proliferative DR (PDR).

In this paper, Sect. 2 gives a brief review of the related works on DR. Section 4 presents the proposed ISDR. Section 5 discusses the result obtained by ISDR. Finally, Sect. 6 draws some conclusions and focuses on future works related to ISDR.

2 Literature Survey

Automated analysis and classification of the retina as DR or non-DR using two-field mydriatic fundus photography has been proposed in [3]. Detection methods and processes such as multi-level wavelet decomposition, recursive region growing algorithm, histogram analysis, intensity transformations, etc., have been performed. The method obtains a sensitivity of 80% and a specificity of 50%. Rajalakshmi et al. [4] proposed an artificial intelligence-based automated software named EyeArtTM, for the detection of diabetic retinopathy and sight-threatening diabetic retinopathy (STDR) on fundus images. The EyeArtTM AI algorithm provided the diagnosis for retinal images of 296 patients. DR was detected by the ophthalmologists in 191 (64.5%) and by the AI software in 203 (68.6%) patients. STDR was detected in 112 and 146 patients, which accounts for 37.8% and 49.3%, respectively. The AI software showed 95.8%, i.e., 95% confidence interval 92.9–98.7 sensitivity and 80.2%, i.e., 95% confidence interval 72.6–87.8 specificity for detecting any DR and 99.1%, i.e., 95% confidence interval 95.1–99.9 sensitivity and 80.4%, i.e., 95% confidence interval 73.9–85.9 specificity in detecting STDR.

Ege et al. [5] used the Bayesian classifier, the Mahalanobis distance classifier, and the K-NN classifier to detect microaneurysms, hemorrhages, exudates and cotton wool spots on 134 retinal images. The Mahalanobis classifier performs better and detects microaneurysms, hemorrhages, exudates and cotton wool spots with a sensitivity of 69%, 83%, 99% and 80%, respectively. Fadzil et al. [6] investigated FAZ enlargement with DR progression using a Gaussian Bayes classifier and implemented a DR grading protocol using color fundus images.

Decenciere et al. [7] presented a prototype for the automatic detection of normal examinations on a teleophthalmology network for diabetic retinopathy screening. Rakhlin [8] proposed a deep convolutional neural network (CNN) in the field of medical imaging for the diagnosis of eye fundus images to detect referable DR(r-DR) in Kaggle and Messidor-2 datasets. The model achieves a sensitivity of 99%, specificity of 71% and area under the operator receiving characteristic curve (AUC) of 0.97. The features extracted are drusen, exudates, microaneurysms and cotton wool spots. Li et al. [9] studied on the principles and recent developments made in detection of DR and the emerging trends in the framework of automated analysis of DR images. Fadzil et al. [10] proposed a computerized DR grading system to measure the FAZ. The model uses the FINDeRS database and applies a V-fold cross-validation method to evaluate the performance of the DR system. The system achieves a sensitivity of 84%, specificity of 97% and an accuracy of 95% for all DR stages. Walter et al. [11] proposed a new algorithm for the detection of exudates. The model detects the optic disk using morphological filtering techniques and watershed transformation. The algorithm obtains a mean sensitivity of 92.8% and a mean predictive value of 92.4%.

3 Important Terminologies

Some of the important terminologies that have been associated with the development of the intelligent system for the identification of DR are—foveal avascular zone (FAZ), exudates (EX), microaneurysms (MA), hemorrhages (HE), cotton wool spots (CWS), intra-retinal microvascular abnormalities (IRMA) and many more [3, 5, 6,

10–12]. Here, the proposed system considers only two main retinal lesions of DR, i.e., FAZ and MA. FAZ is an area devoid of any blood capillaries and found in the macula region of the eyes [10]. In healthy patients, the diameter size of the FAZ is approximately 500 μm and has an area of approximately 0.4 mm^2. The area of the FAZ enlarges in patients suffering from diabetic retinopathy [12–14]. Microaneurysms are identified by ophthalmoscopy as deep-red dots having a range of 25–100 μm in diameter. They are usually the first visible sign of diabetic retinopathy. Thus, the detection and prediction of microaneurysms in the fundus images are important in the process of predicting the severity of the disease.

4 Proposed Intelligent System for Diabetic Retinopathy (ISDR)

The proposed ISDR begins with the retinal image dataset taken from the kaggle.com repository [8]. There have been many other datasets such as Indian Diabetic Retinopathy Image Dataset (IDRiD) [9], e-ophtha [7, 9], FINDeRS (yet to be approved) [9, 10], messidor [8] which have been very new. However, Kaggle dataset has been found to be convenient and accurate for the purpose of validation and hence has been chosen for the same. The initial dataset contains 30000 retinal fundus images. These images are labeled into five categories—no DR—Grade 0, mild non-proliferative DR—Grade 1, moderate NPDR—Grade 2, severe NPDR—Grade 3 and proliferative DR—Grade 4. Figure 1 presents the classification of the grading classes in the dataset.

The dataset is highly skewed as seen in Fig. 2. This skewness needs to be removed such that the classifier associated with the system does not make a biased decision toward a particular category. Hence, to remove the excess data, the proposed ISDR considers only 500 images in each category resulting in a total of 2500 images. Figure 3 shows the layout of the proposed ISDR.

All the computations and analysis are done on this preprocessed dataset. Image enhancement is performed to improve the quality of the image, such as contrast, sharpness, intensity, brightness and denoising, such that fine details can be obtained

Fig. 1 Classes in the dataset

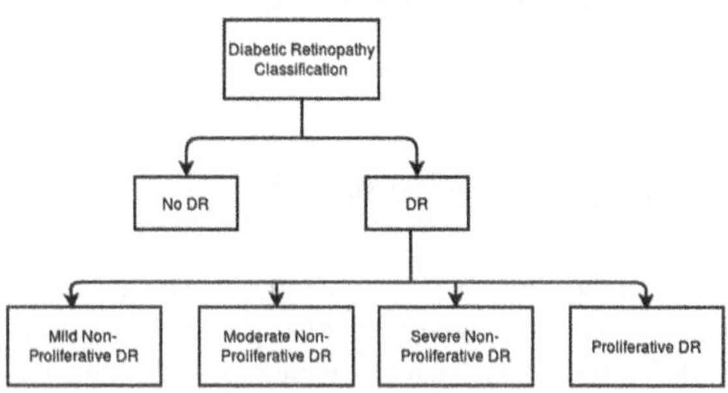

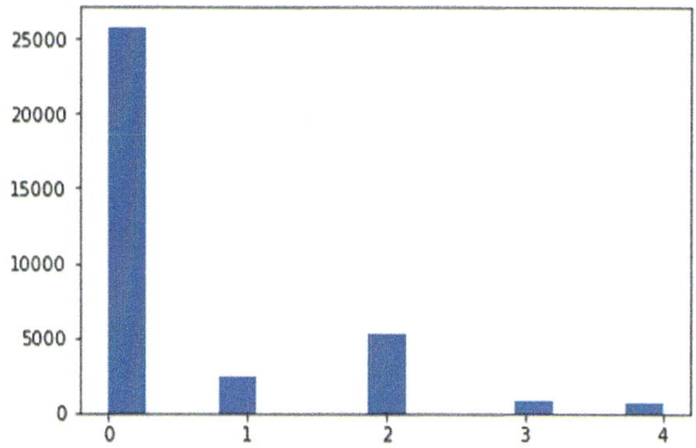

Fig. 2 Histogram analysis of dataset 1

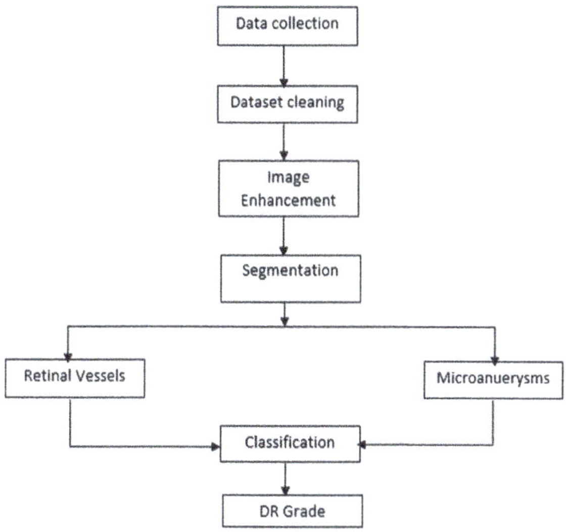

Fig. 3 Proposed ISDR

from the images and then further processing is done on them. Image segmentation techniques are used to segregate blood vessels from the background and to count MA. Then SVM classifier is used to classify the images into different categories.

4.1 Image Enhancement

During the image enhancement stage, the ISDR lays focus on enhancing the contrast of the images to extract information that exists within a low dynamic range of that gray level image. The aim of image enhancement is to improve the interpretability of information in images for human graders and to provide better input for other automated image processing techniques. The ISDR performs image enhancement

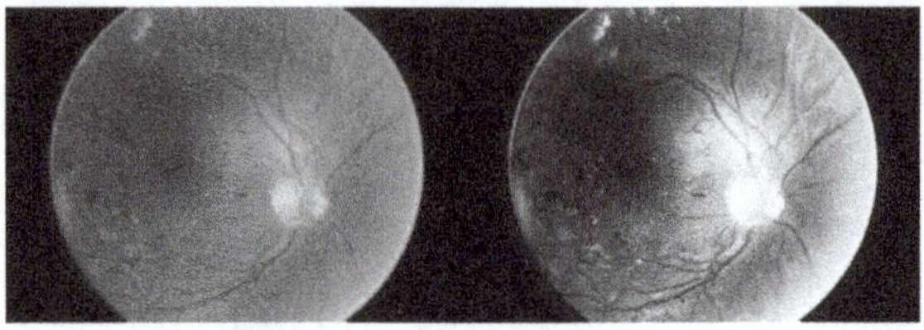

Fig. 4 Original image versus enhanced image

in order to make the blood vessels and MA more distinct than the background. Contrast limited adaptive histogram equalization (CLAHE) algorithm is used for image enhancement. CLAHE is a computer image processing technique and is used to enhance the contrast of the image.

In the histogram of the original image, it can be seen that the histogram seems continuous for certain ranges of gray levels. But in the histogram of the contrast-enhanced image, the graph seems mostly discrete. This is done by the CLAHE technique in order to increase the distance in gray levels between pixels having different gray level values. This makes them easily distinguishable from each other, thereby increasing the contrast between the different pixels.

Figure 4 shows the enhanced image that is obtained after applying CLAHE on the original image. Figure 5a, b describes the histogram of the pixel distribution of the original image and the enhanced image, respectively.

4.2 Image Segmentation

Before applying the classifier on the dataset, it is important to segment the important features from the image. The features that are considered in the system are FAZ and MA. To obtain the FAZ area, it is firstly necessary to segment the blood vessels. Once the FAZ area is calculated, MAs are segmented to obtain the total count of it. The

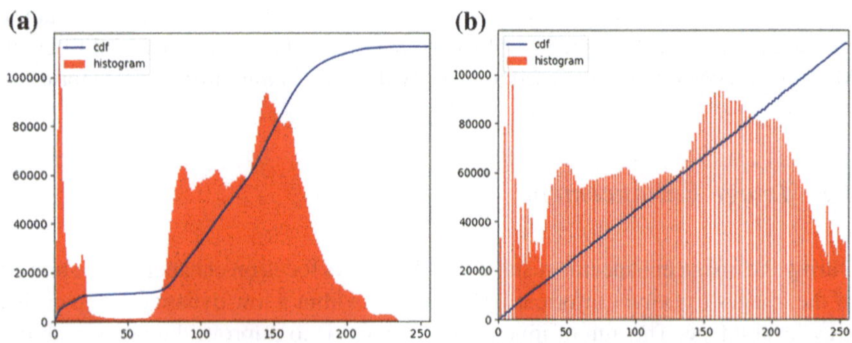

Fig. 5 **a** Histogram analysis of original image, **b** histogram analysis of enhanced image

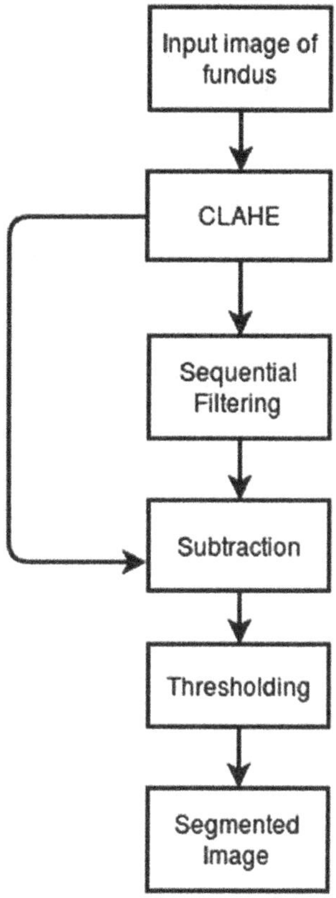

Fig. 6 Blood vessel segmentation flowchart

segmentation of the MA is done using the Canny edge detection algorithm (CEDA) and blood vessel segmentation is done using adaptive sequential filtering (ASF).

4.2.1 Blood Vessels Segmentation

ASF is used to extract blood vessels from the image, which gives another image with an average intensity of each region applied over it. This enhances and restores the data by removing noise without causing any significant blur in the image. Later on, this image is subtracted from the output of CLAHE. The flowchart given below in Fig. 6 illustrates the process of segmenting blood vessels.

4.2.2 FAZ Area Calculation

The FAZ area is calculated by connecting the 12 closely situated endpoints of the blood vessels from the center of the macula region and by obtaining the pixel count of the area enclosed within the macula. This requires a clear segmented image of the blood vessels of the eye. Hence, the segmentation of the blood vessels is a crucial step in calculating FAZ area.

4.2.3 Microaneurysms Segmentation

The green channel of the image is extracted as it gives the best contrast between the microaneurysms and other bright parts such as optic disk and exudates. The image contrast is further stretched by applying adaptive histogram equalization before using edge detection technique (Canny method) to detect the outlines of the image. The boundary is detected by filling up the holes and a disk-shaped structuring element (SE) of radius 6 is created with morphological opening operations, i.e., erosion and dilation [11]. The edge detection image is then subtracted from the image with boundaries to obtain an image without boundaries. Then, the holes or gaps are filled, resulting in microaneurysms and other unwanted artifacts. The blood vessels which are detected using the above-mentioned method are subtracted from the image of microaneurysms and artifacts. The flowchart of the segmentation of microaneurysms is shown in Fig. 7.

4.2.4 Canny Edge Detection

Canny edge detection performs noise reduction, gradient calculation, non-maximum suppression, double threshold and edge tracking by hysteresis. During noise reduction, Gaussian blur is applied to smoothen the edges as they are highly sensitive to noise. The image convolution technique is applied with a Gaussian Kernel (3×3) depending on the expected blurring effect. The gradient calculation detects the edge intensity and direction by calculating the gradient of the image using edge detection operators. Edges correspond to a change of pixels' intensity. Non-maximum suppression is performed to thin out the edges. The purpose of the algorithm is to check if the pixels in the same direction are more or less intense than the ones being processed. In double threshold, three kinds of pixels-strong, weak and non-relevant are identified.

Strong pixels have a high intensity to contribute to the final edge. Weak pixels are neither strong nor non-relevant for edge detection. Other pixels are considered as non-relevant for the edge. Edge tracking by hysteresis transforms weak pixels into strong ones, if and only if at least one of the pixels around the one being processed is a strong one. After segmentation, the extraction of two features, namely the FAZ area and microaneurysms count, is carried out. The values of these two features are used for image classification.

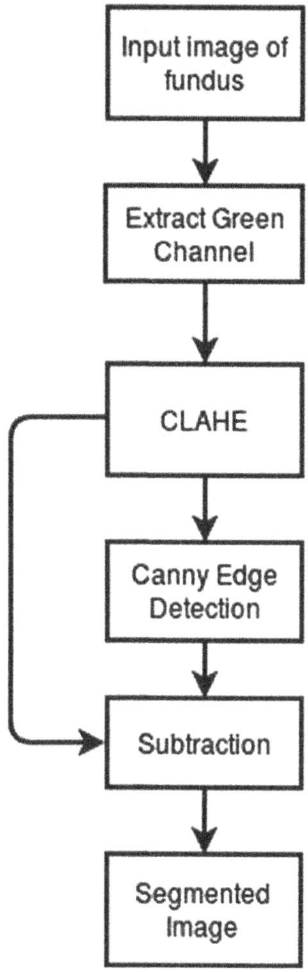

Fig. 7 Microaneurysms segmentation flowchart

4.3 Classification

Three classifiers are used by the proposed system. They are described briefly below.

4.3.1 Support Vector Machine (SVM)

SVM is a classifier that distinguishes categories formally based on a separating hyperplane. In other words, given a labeled training dataset, SVM creates an optimal hyperplane that categorizes the data into different classes. SVM is used as the FAZ area and microaneurysms are nonlinear in nature.

4.3.2 K-NN

The K-NN is a classifier that classifies a data point based on the proximity of the data point to K-nearest neighbors of the data point. It is classified in the class that has the most number of points in the K-neighbors. The proposed ISDR considers the value of K as 5.

4.3.3 Gaussian Naïve Bayes (GNB)

Gaussian Naïve Bayes is a special Naïve Bayes algorithm used when the features have continuous values in the form of a Gaussian distribution, for predictive modeling. This Gaussian distribution is done for the extension of real-valued attributes. It is one of the easiest forms of estimating the distribution of the data, by estimating the mean and standard deviation from the training data.

5 Results and Discussions

2500 images of Kaggle dataset are taken equally from all the classes. The 80% of data is used as training set and 20% as testing set for all the evaluations.

5.1 Pre-processing

This section presents the results that are obtained during the evaluation of the proposed system. The results are evaluated on all the 2500 images. Figure 8 shows a sample result of the CLAHE applied to the images.

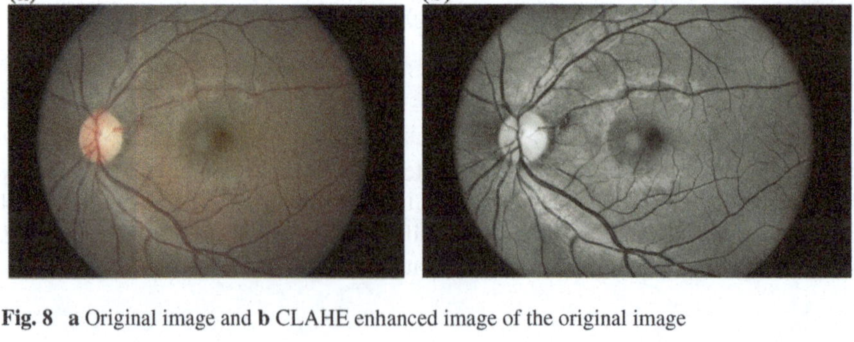

Fig. 8 **a** Original image and **b** CLAHE enhanced image of the original image

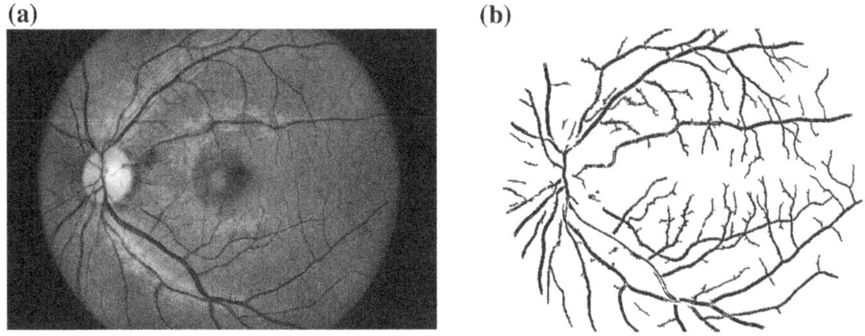

Fig. 9 **a** Enhanced image and **b** blood vessel segmentation on the enhanced image

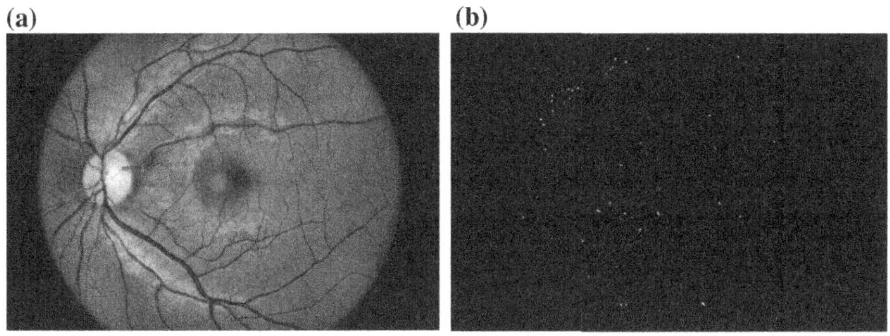

Fig. 10 **a** Enhanced image, **b** microaneurysms segmentation on the enhanced image

Figures 9 and 10 show the result obtained after the segmentation of the blood vessels and the result obtained after the segmentation of the microaneurysms, respectively, from the CLAHE enhanced image.

5.2 Classification

FAZ area and microaneurysms count are fed into different classifiers to build the proposed ISDR. Table 1 represents the performance of the various classifiers. Table 1 also shows the performance when FAZ area and microaneurysms are taken individually, and together. From the experimental results, it is clear that when both features are taken together performance increases. SVM classifier performs better than all other classifiers due to its nonlinear behavior.

Table 2 represents the precision levels of the different classifiers according to the severity levels. 0-sev represents the category with 0 severity level, 1-sev that of 1 severity level, 2-sev that of 2 severity level, 3-sev that of 3 severity level and 4-sev that of 4 severity level. Precision is calculated using FAZ and MA as it gives better

Table 1 Accuracy of different models

Classifier	Features taken	Accuracy (%)
Gaussian Naive Bayes	FAZ area	35.4
K-NN	FAZ area	35.6
K-NN	Microaneurysms	41.6
K-NN	FAZ, microaneurysms	51.4
SVM	FAZ area	36
SVM	microaneurysms	41.4
SVM	FAZ, microaneurysms	51.8

Table 2 Precision of different severity levels of different models

Classifier	0-sev (%)	1-sev (%)	2-sev (%)	3-sev (%)	4-sev (%)
Gaussian Naive Bayes	66.3	19.19	31.91	11.65	48.62
K-NN	61.05	51.51	42.42	48.54	51.37
SVM	81.05	39.39	40.42	59.22	40.36

performance. It is observed from the experimental results that SVM performs better in most of the levels. It is noted here that FAZ and MA give a better prediction, especially for severity level 0. It can be concluded from this that FAZ and MA are early sign of the onset of diabetic retinopathy.

Similarly, Table 3 represents the recall levels of the different classifiers according to the severity levels. It is observed from the experimental results that ISDR with SVM and K-NN produce approximately uniform recall rate. It should be noted here that recall rate of four severity level is higher than other severity levels.

Table 4 shows F1-score values by different classifiers. It can be observed from the results that K-NN and SVM produces approximately uniform F1-score values.

Table 3 Recall of different severity levels by different classifiers

Classifier	0-sev (%)	1-sev (%)	2-sev (%)	3-sev (%)	4-sev (%)
Gaussian Naive Bayes	42	25.33	26.62	23.6	53
K-NN	53.21	42.5	48.27	50.5	65.8
SVM	51.67	41.93	49.35	49.19	77.19

Table 4 F1-measure of different severity levels of different models

Classifier	0-sev (%)	1-sev (%)	2-sev (%)	3-sev (%)	4-sev (%)
Gaussian Naive Bayes	51.33	21.59	28.28	14.88	50.37
K-NN	56.86	46.57	45.15	49.5	57.65
SVM	63.11	40.62	44.44	53.74	53

6 Conclusion and Future Works

This paper proposes a system called ISDR for early detection of DR using SVM, Gaussian NB and K-NN classifiers. Image enhancement is done to enhance the contrast of the images to extract information that exists within a low dynamic range of that gray level image. Image segmentation is employed to segment the important features from the image. Finally, SVM, Gaussian NB and K-NN classifiers are used to classify an image into five categories namely, no DR—Grade 0, mild nonproliferative DR—Grade 1, moderate NPDR—Grade 2, severe NPDR—Grade 3 and proliferative DR—Grade 4 for early detection of DR. Experimental results show that SVM performs better than all other classifiers. It should be noted here that classification accuracy suffers in the dataset primarily due to the segmentation process. If the segmentation of the features could be done properly, a better result could be obtained. Also, more features can be explored to produce better results. Features such as exudates, hemorrhages, etc., can be used along with the FAZ area and the microaneurysms count to produce better results.

There is a limited number of professionals for the diagnosis of DR. This presents a reason for us to automate the process. The society can be made better by the use of computers and this paper helps in realizing this fact. Finally, a Web-based application will make the process extremely fast as it can be accessed from anywhere.

References

1. D.S. Fong, L. Aiello, T.W. Gardner, G.L. King, G. Blankenship, J.D. Cavallerano, F.L. Ferris, R. Klein, Retinopathy in diabetes. Diab. Care **27**(1), 84–87 (2004)
2. J.K.H. Goh, C.Y. Cheung, S.S. Sim, P.C. Tan, G.S.W. Tan, T.Y. Wong, Retinal imaging techniques for diabetic retinopathy screening. J. Diab. Sci. Technol. **10**(2), 282–294 (2016)
3. P.S. Kumar, R.U. Deepak, A. Sathar, V. Sahasranamam, R.R. Kumar, Automated detection system for diabetic retinopathy using two field fundus photography. Proc. Comput. Sci. **93**, 486–494 (2016)
4. R. Rajalakshmi, R. Subashini, R.M. Anjana, V. Mohan, Automated diabetic retinopathy detection in smartphone-based fundus photography using artificial intelligence. Eye **32**, 1138–1144 (2018)
5. B.M. Ege, O.K. Hejlesen, O.V. Larsen, K. Møller, B. Jennings, D. Kerr, D.A. Cavan, Screening for diabetic retinopathy using computer based image analysis and statistical classification. Comput. Methods Programs Biomed. **62**(3), 165–175 (2000)
6. M.H.A. Fadzil, N.F. Ngah, T.M. George, L.I. Izhar, H. Nugroho, H.A. Nugroho, Analysis of foveal avascular zone in colour fundus images for grading of diabetic retinopathy severity, in *Annual International Conference of the IEEE Engineering in Medicine and Biology*, pp. 5632–5635, Aug 2010
7. E. Decencière, G. Cazuguel, X. Zhang, G. Thibault, J.C. Klein, F. Meyer, B. Marcotegui, G. Quellec, M. Lamard, R. Danno, D. Elie, TeleOphta: Machine learning and image processing methods for teleophthalmology. IRBM **34**(2), 196–203 (2013)
8. A. Rakhlin, Diabetic Retinopathy detection through integration of Deep Learning classification framework. bioRxiv 1–11 (2013)
9. B. Li, H.K. Li, Automated analysis of diabetic retinopathy images: principles, recent developments, and emerging trends. Curr. Diab. Rep. **13**(4), 453–459 (2013)

10. M.A. Fadzil, L.I. Izhar, H. Nugroho, H.A. Nugroho, Analysis of retinal fundus images for grading of diabetic retinopathy severity. Med. Biol. Eng. Compu. **49**(6), 693–700 (2011)
11. T. Walter, J.C. Klein, P. Massin, A. Erginay, A contribution of image processing to the diagnosis of diabetic retinopathy-detection of exudates in color fundus images of the human retina. IEEE Trans. Med. Imaging **21**(10), 1236–1243 (2002)
12. A. Sopharak, B. Uyyanonvara, S. Barman, Automatic exudate detection from non-dilated diabetic retinopathy retinal images using fuzzy c-means clustering, sensors **9**(3), 2148–2161 (2009)
13. S. Roychowdhury, D.D. Koozekanani, K.K. Parhi, DREAM: diabetic retinopathy analysis using machine learning. IEEE J. Biomed. Health Inform. **18**(5), 1717–1728 (2013)
14. X. Xia, C. Xu, B. Nan, Inception-v3 for flower classification, in 2nd international conference on image, vision and computing (ICIVC), pp. 783–787 (2017)

Markov Chain Models for the Near Real-Time Forecasting of Australian Football League Match Outcomes

Casey Josman, Ritu Gupta, and Sam Robertson

Abstract The Australian Football League (AFL) is a fast-paced invasion style ball game where two teams of 18 players jockey for a position through the strategic use of on-field tactics and skills. Each match consists of four quarters which range in length from 20 to 40 min and are heavily influenced by the irregular nature of game stoppages. The purpose of this study was to identify whether it was possible to model and subsequently forecast match outcome based on both prior match information as well as real-time transactional data. 23 professional level matches were analysed with each match containing an average of 1907 transactional epochs. To account for the irregular frequency at which each epoch occurs, a time-inhomogeneous Markov process was adopted, with model evaluation assessed on both epoch accuracy and final match result. The model performed notably well with an average epoch accuracies in excess of 71% and match outcome results in excess of 78%. The results of this study demonstrate that accurate near real-time prediction is achievable under real-world conditions using real match AFL transactional data.

Keywords Australian Football League · AFL · Forecast · Markov model · Multinomial logistic regression

1 Introduction

This paper aims to outline an approach for near real-time forecasting of Australian Football League (AFL) match outcomes utilising a Markov chain model. The purpose of this research is twofold; firstly, to understand the underlying forces of team

C. Josman (✉) · R. Gupta
Department of Mathematics and Statistics, Curtin University, Bentley, WA 6062, Australia
e-mail: Casey.Josman@Curtin.edu.au

S. Robertson
Institute of Sport Exercise and Active Living, Victoria University,
Footscray, VIC 3011, Australia

© Springer Nature Singapore Pte Ltd. 2020

A. K. Nagar et al. (eds.), *Soft Computing for Problem Solving 2019*,
Advances in Intelligent Systems and Computing 1139,
https://doi.org/10.1007/978-981-15-3287-0_9

momentum and performance and their effect on match outcomes, and secondly, to allow coaches to make informed tactical decisions based on the aforementioned forces. The rationale behind this is to give coaches and training staff a deeper understanding of the forces that govern victory and allow them to harness it in near real time, evaluating outcome probabilities alongside traditional metrics of team performance. This has the potential to enable teams to change tactics and pre-empt substitutions on the fly resulting in a more strategic and hence action-packed game.

AFL is a fast-paced invasion style game with quick transitions between play and stoppage, and no hard limit to match time, wherein two teams of 18 compete on an oval-shaped field with each team's objective being to score more points than their opponent. Points are scored by either kicking the ball between the centre goal posts—scoring a goal and being awarded 6 points, or by kicking the ball between either of the two outer (behind) posts—scoring a behind and being awarded 1 point [1].

A Markov chain model was chosen due to its ability to handle large data sets and its newly found prevalence within the literature. Building on from the work of Shirley [2] who used the Markov model with limited data to forecast basketball results for a given team against an average opponent, Štrumbelj and Vračar [3] set out to determine whether it was possible to simulate a basketball game between two specific teams. In order to do this, they employed a possession-based Markov model which was evaluated against three separate forecasting approaches; logistic regression, latent strength ranking model and bookmaker odds. From these comparisons Štrumbelj and Vračar [3] found that whilst their Markov model whilst successful (68.72% accuracy) it was less accurate than the bookmaker odds model (70.42% accuracy), and was even improved on when using a combination of the Markov model and Elo model (69.72% accuracy) thus showing that there is still room for improvement.

Conversely, a multinomial logistic regression model for initial probabilities was chosen due to its wide use and favourable results in various sporting scenarios. Notably, Akhtar and Scarf [4] developed a multi-stage (pseudo-real time) prediction strategy for in-play test cricket outcomes by predicting the outcome of a five-day international test match using individual model variations for each inning of play. Unsurprisingly, accuracies at the start of the first day (based solely on static data) yielded an accuracy of 59.6%, whilst by tea time on the fifth day, they achieved a prediction accuracy of 95.2%.

This paper is structured as follows. Section 2 describes the statistical framework used to forecast AFL match outcomes in near real time. Section 3 collates the results from each of the two initial probability sub-models. Section 4 presents a discussion of the model results, evaluating model performance and determining whether the two models converge to stationarity. Finally, Sect. 5 presents our conclusions.

2 Methods

The objective of the Markov model methodology outlined within this paper is the accurate calculation of near real-time outcome probabilities at points of interest during a match. These points are canonically known as transaction events and represent any noteworthy on-field incident, for example, when a player kicks the ball into the opposing team's half of the field. For the purposes of this paper, these transaction events will be referred to as epochs and focus primarily on commonly accepted team performance indicators [5], such as score differentials (margin with respect to the home team), behinds and kicks. This section, therefore, provides an overview of the methodology presented in Fig. 1 and includes data acquisition and processing, data analysis and results generation.

2.1 Data Acquisition

A major factor in any mathematical model is the quality of data used for both model creation and testing. With the issue of big data and its widespread adoption within the sporting world, it is important that heavy scrutiny be placed upon data prior to its use. The two types of data utilised for this research can be summarised as follows: static data (known prior to the match) which is widely accessible and can be found on a myriad of online repositories, and dynamic data (gathered during the match) which is restricted to AFL teams and the companies that gather said data. Due to the proprietary nature of the aforementioned dynamic data, the final model has been restricted to matches played by the Western Bulldogs[1] during the 2015 AFL season.

At its core, the data collected is structured as follows; for a given match m between home team i and away team j, a set of feature data $F_t = \{S, D_t\}$ is computed where $S = \{S_i, S_j\}$ and $D_t = \{D_{(i,t)}, D_{(j,t)}\}$ are the sets of static and dynamic data at time t, respectively, and these data sets have been summarised in Table 1.

2.2 Data Processing

2.2.1 Static Data

Static data was collected from AFL Tables [6] and contained information pertaining to all 390 matches played by the Western Bulldogs during the 2001–2015 seasons. This data was pre-processed such that the home and away team allocations provide an unbiased derivation of the home and away team assignments for each match [7]. Guided by the research of Reicher [7], this methodology removes both inter- and intra-club biases introduced through economic and political agendas, and instead

[1]The Western Bulldogs is a professional AFL team headquartered in the Australian state of Victoria.

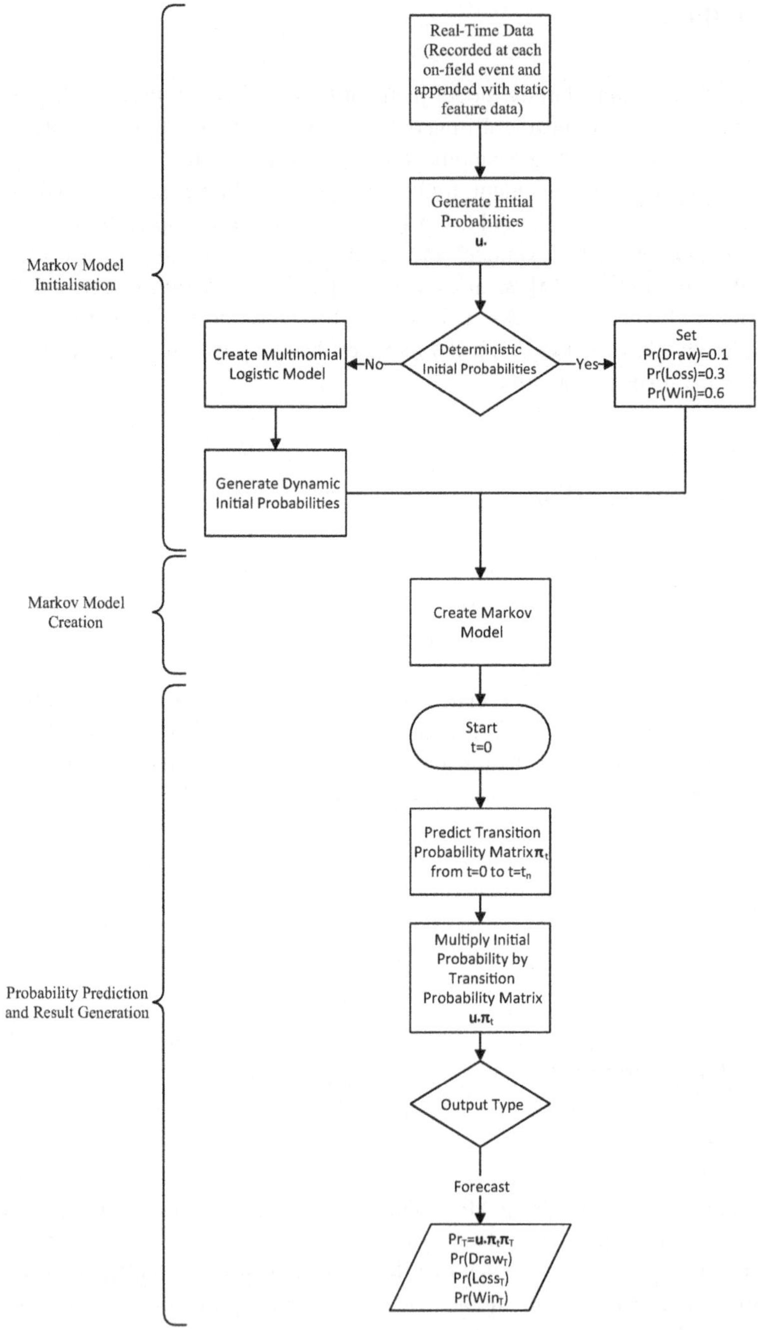

Fig. 1 An overview of the applied Markov model

Table 1 Data summary per data type

	Static data	Dynamic data
Season	2001–2015	2015
Matches	390	23
Features	6	9
Average number of epochs per match	1	1907

relies on stadium conditions (location and crowd composition) which have been shown to directly affect player performance.

As such, for any pairing of teams $(i, j) \in (\mathscr{H}, \mathscr{A})$, in a given match m, the home and away teams are defined as follows: if either team i or j are playing at their home ground, then assign the home team accordingly, and however, if both teams i and j share the same home ground or are both playing an away game, then assign the home team to that team which has the highest official membership number. This revised team assignment yields a home team win probability of 0.607[2] which whilst slightly higher than average holds with the paradigm of home advantage outlined by Stefani and Clarke [8] and Clarke [9]. Additional statistics relating to team form and performance (Head2Head,[3] PastHome,[4] and PastAway[5]) were then calculated (Eqs. 1–3) such that for a given match m between home and away teams i and j, the result $R_{(\mathscr{H},\mathscr{A},m)} = 1$ if team $\mathscr{H}$ wins and $R_{(\mathscr{H},\mathscr{A},m)} = 0$ if team $\mathscr{H}$ loses.

$$\frac{\sum_{g=m-6}^{m-1} \left(R_{i,j,g}\right)}{6} \tag{1}$$

$$\frac{\sum_{a=1}^{5} \sum_{g=m-5}^{m-1} \left(R_{i,a,g}\right)}{5} \tag{2}$$

$$\frac{\sum_{a=1}^{5} \sum_{g=m-5}^{m-1} \left(R_{j,a,g}\right)}{5} \tag{3}$$

[2]The recalculated probability of a home team winning any given match after the home and away team reallocation described above.

[3]The percentage of games over the past 6 games for which the home team i has won against the away team j.

[4]The percentage of games over the past 5 games for which the home team i has won against any opponent a.

[5]The percentage of games over the past 5 games for which the away team j has won against any opponent a.

2.2.2 Dynamic Data

Dynamic data was provided by Champion Data and contained all epoch information pertaining to the 23 matches played by the Western Bulldogs during the 2015 season. This data was extracted from an Extensible Mark-up Language (XML) feed using an algorithm written in R script [10] and processed to remove extraneous epoch data. The remaining epochs were then extracted and tabulated in a contingency table for each observed epoch. Real-time statistics (Margin,[6] and ResT[7]) were also calculated as follows (Eqs. 4–5);

$$\text{Margin} = \text{Score}_i - \text{Score}_j \tag{4}$$

$$\text{ResT} = \begin{cases} \text{Win} & \text{Margin} > 0 \\ \text{Loss} & \text{Margin} < 0 \\ \text{Draw} & \text{Margin} = 0 \end{cases} \tag{5}$$

2.2.3 Overview of Model Features

As per the static and dynamic features described above, the final model makes use of the following features (Table 2).

2.3 Data Analysis

Dynamic stochastic processes such as Markov models have begun recently appearing as alternatives to more traditional forecasting methodologies, with Markov model approaches to prediction and forecasting fast becoming a preferred method for sports analysts due to their ability to accurately model and simulate complex state transitions without relying on time-homogeneous event observations [11]. As per Fig. 1, the Markov model is initialised and evaluated in three distinct stages: stage 1: Markov model initialisation; stage 2: Markov model creation; and stage 3: probability forecast and results generation.

This research in its implementation is concerned with two separate statistical models: firstly, the multinomial logistic regression model which is used in combination with static data to generate initial starting probabilities for the non-deterministic Markov chain model variant, and secondly, the Markov Chain model which is responsible for calculating the transition probability matrices (Eq. 9) which are fed into the model's forecast generating function (Eq. 30).

[6]The score margin with respect to the home team i at time t.

[7]The result with respect to the home team i at time t.

Table 2 Model feature summary

Static features	Home rank	Away rank	Past home	Past away	Head2Head	Round			
Dynamic features	Behind (Home)	Rush behind (Home)	Kick in (Home)	Kick effective (Home)	Behind (Away)	Rush behind (Away)	Kick in (Away)	Kick effective (Away)	Margin

2.3.1 Multinomial Logistic Regression Model

Multinomial logistic regression is a generalised logistic regression model that expands the bimodal predictive capabilities of a logistic regression and applies it to multi-class problems [12]. The basic formulation for a logistic model is given by Eq. 6 which can be generalised by Eq. 7

$$f\,(\Pr\,(S)) = \ln\left(\frac{\Pr\,(S)}{1-\Pr\,(S)}\right) = \beta_0 + \beta_1 x_1 + \beta_2 x_2 + \cdots + +\beta_n x_n \tag{6}$$

where $X \in \{\text{Win, Loss, Draw}\}$ is the desired match outcome, and S is the set of static feature data and

$$\Pr\,(Y = X\,|S) = \frac{e^{\beta_{X,0}+\sum_{v\in S}\beta_{X,v}x_v}}{1+\sum_{r\in R}\left(e^{\beta_{r,0}+\sum_{v\in S}\beta_{r,v}x_v}\right)} \tag{7}$$

where $R = \{\text{Win, Loss, Draw}\}$ is the set of all possible outcomes and 0 is the reference case.

2.3.2 Markov Chain Model

Theoretically, a Markov model is stochastic process structured as a set of successive Markov chains occurring in either discrete or continuous state space. A Markov chain is a sequence of discrete observations satisfying the Markov property

$$\Pr\left(X_{t_{j+1}}|F_1,\ldots,F_{t_j}\right) = \Pr\left(X_{t_{j+1}}|F_{t_j}\right) \tag{8}$$

such that $X_t = \{D, L, W\}$ is the state space, and $F_{t_j} = \left\{S, D_{t_j}\right\}$ is the set of static and dynamic features observed during an AFL match and

$$\Pr\left(X_{t_{j+1}}|F_{t_j}\right) = \begin{bmatrix} p_{11} & \cdots & p_{13} \\ \vdots & \ddots & \vdots \\ p_{31} & \cdots & p_{33} \end{bmatrix} = \Pr\left(t_j, t_{j+1}\right) = e^{\left(Q\left(F_{t_j}\right)t_j\right)} \tag{9}$$

is the probability of observing an outcome X at time t_{j+1} given observed feature data F up to and including time t_j where

$$Q\left(F_{t_j}\right) = \begin{bmatrix} q_{11} & \cdots & q_{13} \\ \vdots & \ddots & \vdots \\ q_{31} & \cdots & q_{33} \end{bmatrix} \tag{10}$$

is the transition intensity matrix after observing feature data F up to and including time t_j [13], which is solved using the Kolmogorov forward equation making use of partial differential equations and eigenvalue decomposition to solve for each $q_{s(t_j)s(t_{j+1})}$ [14].

$$\begin{aligned} \frac{\partial \Pr(t)}{\partial t} &= \Pr(t) \\ \Rightarrow \frac{1}{\Pr(t)}\frac{\partial \Pr(t)}{\partial t} &= Q \\ \Rightarrow \frac{\partial \ln(\Pr(t))}{\partial t} &= Q \\ \Rightarrow \int \ln(\Pr(t)) &= \int Q\partial t \\ \Rightarrow \Pr(t) &= e^{(Qt)} \\ \Rightarrow \Pr(t) &= ADA^{-1} \end{aligned} \tag{11}$$

The decomposition in Eq. 11 is such that D is a diagonal matrix with element $\{(i,j) : i = j\}$ corresponding to the exponential of the ith distinct eigenvalue of Q and A is a matrix with eigenvectors corresponding to the aforementioned eigenvalues. Hence, to forecast $\Pr(t_1, t_2)$, that is, the probability transition matrix from t_1 to t_2, the decomposition $AD^{t_2-t_1}A^{-1}$ is derived and solved. In addition to this, the values for the transition matrix Q are found by maximising the likelihood function $L(Q|\theta)$ for each of the unknown parameter values $\theta = \left\{ q_{\left(S_{t_j}\right)\left(S_{t_j+1}\right)}, \beta_{\left(S_{t_j}\right)\left(S_{t_j+1}\right)} | \left\{S_{t_j}, S_{t_j}+1\right\} \in X_t \right\}$ [15]. The likelihood described above can be expressed by Eqs. 12 and 13 where successive states $S\left(t_j\right)$ and $S\left(t_{j+1}\right)$ occur at times t_j and t_{j+1} for an index i which contains the set of all observable matches $M = \{1, 2, \ldots, m\}$

$$L(Q|\theta) = \prod_{i=1}^{M} L_{i,j} \tag{12}$$

$$L_{i,j} = p_{S(t_j)S(t_{j+1})}\left(t_{j+1} - t_j|\theta\right) \tag{13}$$

Equivalently, once the values for Q have been found, it is possible to calculate $P(t)$ by using the matrix exponential $\exp(Qt)$ such that

$$\exp(Qt) = \sum_{k=0}^{\infty} \frac{1}{k!}(Qt)^k \tag{14}$$

with $(Qt)^k = Qt \times Qt \times \cdots \times Qt$. However, whilst this formulation is simple enough, it does not allow for the inclusion of time-varying covariates, and this is achieved by altering the above formulation as follows [15].

$$q_{\left(S_{t_j}\right)\left(S_{t_j+1}\right)}\left(t_j\right) = q^0 e^{\beta^T_{\left(S_{t_j}\right)\left(S_{t_j+1}\right)} z(t_j)} \tag{15}$$

$$L_{i,j} = e^{q_{S(t_j)S(t_j)}\left(t_{j+1} - t_j|\theta\right)} q_{S(t_j)S(t_{j+1})} \tag{16}$$

where $z\left(t_j\right) = Cov_{t_j} - Cov_{Mean}$ is the difference between the observed covariate values at time t_j and the mean model values, and q^0 is the log-baseline estimate of $Q\left(t_j\right)$. This enables the approximation of

$$Q\left(t_j\right) = e^{\sum q^0 + q^0_{S\left(t_j\right)} z\left(t_j\right)} \tag{17}$$

from which it is possible to calculate $P\left(t_1, t_2\right)$ as

$$\Pr\left(t_1, t_2\right) = \Pr\left(t_1, t_2\right) \times \Pr\left(t_2, t_{j-1}\right) \times \Pr\left(t_{j-1}, t_j\right) \tag{18}$$

where the epochs $\left\{t_1, \ldots, t_j\right\}$ are the inhomogeneous time points at which on-field transactions are recorded, and

$$\begin{aligned} \Pr\left(t_1, t_2\right) &= \exp\left(t_2 - t_1\right) Q\left(t_1\right) \\ \Pr\left(t_2, t_{j-1}\right) &= \prod_{t_2}^{t_{j-1}} \exp\left(t_i - t_{i-1}\right) Q\left(t_i\right) \\ \Pr\left(t_{j-1}, t_j\right) &= \exp\left(t_j - t_{j-1}\right) Q\left(t_{j-1}\right) \end{aligned} \tag{19}$$

Stage 1 After data processing, an initial transition matrix is formulated as the percentage of state-specific transitions from state $S\left(t_j\right)$ to $S\left(t_{j+1}\right)$, where $r\left[i\right]$ is the total number of transitions from a specific state $S\left(t_j\right)$ to any state $S\left(t_{j+1}\right)$.

$$\Pr\left(t_0\right) = \begin{array}{cc} & \text{to} \\ & \begin{array}{ccc} \text{Win} & \text{Loss} & \text{Draw} \end{array} \\ \begin{array}{r} \text{from Win} \\ \text{Loss} \\ \text{Draw} \end{array} & \begin{array}{ccc} \frac{(W,W)}{r[1]} & \frac{(W,L)}{r[1]} & \frac{(W,D)}{r[1]} \\ \frac{(L,W)}{r[2]} & \frac{(L,L)}{r[2]} & \frac{(L,D)}{r[2]} \\ \frac{(D,W)}{r[3]} & \frac{(D,L)}{r[3]} & \frac{(D,D)}{r[3]} \end{array} \end{array} \tag{20}$$

In order to assess the stability of our model, two types of initial probabilities were considered: (1) a set of deterministic initial probabilities which conform to the home advantage paradigm [8] $\mathbf{u_d}\left(1\right)$ and (2) a set of non-deterministic initial probabilities $\mathbf{u_s}\left(1\right)$.

$$\mathbf{u_d}\left(1\right) = \left\{\Pr\left(\text{Win}_d\right), \left(\text{Loss}_d\right), \left(\text{Draw}_d\right)\right\} = \left\{0.6, 0.3, 0.1\right\} \tag{21}$$

$$\mathbf{u_s}\left(1\right) = \Pr\left(X_{t_{j+1}} | F_{t_j}\right) = \left\{\Pr\left(\text{Win}_s\right), \Pr\left(\text{Loss}_s\right), \Pr\left(\text{Draw}_s\right)\right\} \tag{22}$$

Stage 2 A Markov chain model was then constructed with match outcome probabilities as a function of both static and dynamic features. This model yields an average Q matrix for the sample of 23 matches

$$Q\left(F_{t_j}\right) = \begin{pmatrix} -0.273 & 0.091 & 0.182 \\ 0.018 & -0.054 & 0.036 \\ 0.000 & 0.006 & -0.006 \end{pmatrix} \tag{23}$$

and hazard ratios $e^{\beta_{S(t_j)S(t_{j+1})}}$ for covariates listed in Table 2.

$$e^{\beta_{DL}} = \{0.979, 1.038, 0.559, 0.440, 0.0821, 1.046, 0.986, 0.812, 1.255, 1.171, 0.850\} \tag{24}$$

$$e^{\beta_{DW}} = \{1.008, 0.985, 2.351, 0.431, 0.463, 1.032, 1.027, 2.084, 0.507, 0.642, 1.530\} \tag{25}$$

$$e^{\beta_{LD}} = \{1.054, 1.028, 0.997, 0.389, 0.020, 1.019, 1.029, 0.871, 0.932, 0.948, 1.067\} \tag{26}$$

$$e^{\beta_{LW}} = \{1.155, 0.978, 1.422, 0.201, 2.225, 0.973, 1.043, 0.840, 1.033, 0.988, 1.028\} \tag{27}$$

$$e^{\beta_{WD}} = \{1.020, 0.951, 2.032, 4.048, 1.926, 1.021, 1.002, 0.628, 1.361, 1.067, 0.870\} \tag{28}$$

$$e^{\beta_{WL}} = \{1.408, 1.086, 0.879, 5.802, 1.047, 1.041, 0.981, 1.008, 0.754, 1.064, 0.960\} \tag{29}$$

Stage 3 At each step of the predictive algorithm, in such a way that each observation takes place at an epoch $t \in [0, \mathbb{T}]$ where t is the current observed time point in a match and $\mathbb{T}$ is the final observed time point in the match. The algorithm produces a forecast of the outcome probabilities at time $\mathbb{T}$ [15]. These results are generated using Eq. 30

$$g\left(F_{t_j}\right) = \mathbf{u}_*\,(1)\Pr\left(t_0, t_j\right)\Pr\left(t_j, \mathbb{T}\right) = \{\Pr\left(\text{Win}_{\mathbb{T}}\right), \Pr\left(\text{Loss}_{\mathbb{T}}\right), \Pr\left(\text{Draw}_{\mathbb{T}}\right)\} \tag{30}$$

2.3.3 Model Evaluation

Each model was evaluated for both epoch prediction accuracy and final result outcome. Epoch prediction accuracy is measured as the percentage of epochs correctly predicted relative to the final outcome at time $\mathbb{T}$, whilst final result outcome is measured as the percentage of forecasts at time $\mathbb{T}$ which match the final result outcome.

All analyses were conducted on a 64-bit Windows operating system computer with Intel® Core™ i7-7700K CPU with 32 GB RAM. All statistical analyses were conducted using R (Version 3.5.1) [10], and making use of the following packages: nnet [16], msm [15], expm [17] and ggplot2 [18].

3 Results

Each model was tested using both deterministic and static initial probabilities and validated using leave-one-out cross-validation, in which at each match a model is trained using the remaining $n - 1$ matches and validated on the left-out case. Results were generated for both epoch prediction accuracy and final result outcome.

Figure 2 is an example of the graphical output generated by the model and plots the forecast outcome probabilities alongside the real-time margin. This particular plot is for the round 13 match between Carlton (home) and Western Bulldogs (away) of the 2015 AFL season. It can be seen that the model is responsive to changes in score margin although it appears to be slightly lagged.

From Table 3 it can be seen that both models perform reasonably well and produce similar results regardless of the initial starting probabilities save for matches 1, 2, 12, 20 and 23 which seem to be outliers. A possibly of this may be due to a lack of transactional granularity or uncharacteristically non-descriptive features.

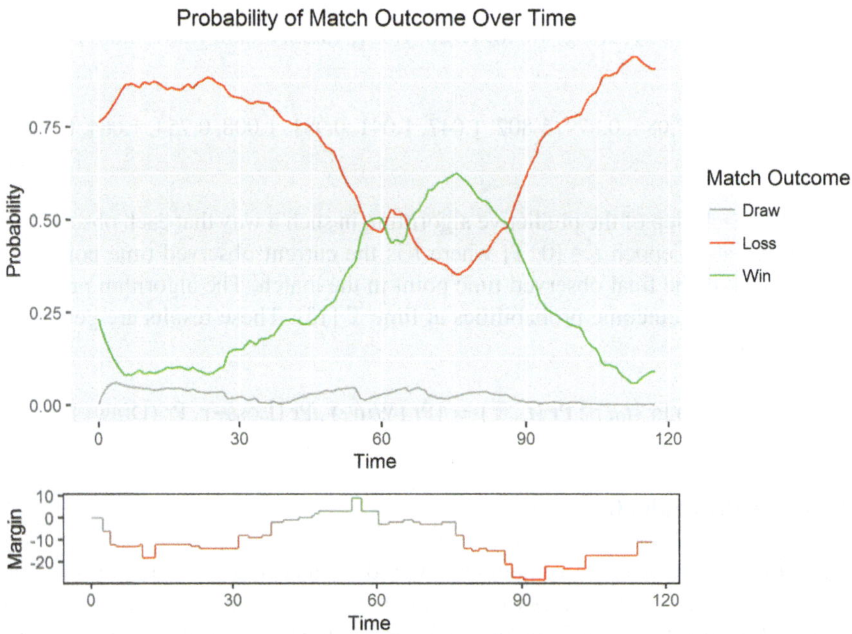

Fig. 2 Forecast outcome probabilities and score margin at time t (in minutes).

Table 3 Model accuracy

Match	Actual result	Deterministic initial probabilities		Static initial probabilities	
		Epoch accuracy[a]	Forecast result[b]	Epoch accuracy[a]	Forecast result[b]
Match 1	Win	0.000	Loss	0.000	Loss
Match 2	Loss	0.121	Win	0.121	Win
Match 3	Win	0.998	Win	0.998	Win
Match 4	Win	0.929	Win	0.953	Win
Match 5	Loss	1.000	Loss	1.000	Loss
Match 6	Loss	0.994	Loss	0.978	Loss
Match 7	Loss	1.000	Loss	0.323	Win
Match 8	Win	0.902	Win	0.843	Win
Match 9	Win	0.995	Win	0.996	Win
Match 10	Win	1.000	Win	1.000	Win
Match 11	Win	1.000	Win	1.000	Win
Match 12	Win	0.051	Loss	0.055	Loss
Match 13	Loss	0.755	Loss	0.775	Loss
Match 14	Win	0.944	Win	0.937	Win
Match 15	Win	0.599	Win	0.609	Win
Match 16	Win	0.974	Win	0.974	Win
Match 17	Loss	1.000	Loss	1.000	Loss
Match 18	Win	0.862	Win	0.862	Win
Match 19	Win	0.997	Win	0.994	Win
Match 20	Win	0.424	Win	0.412	Win
Match 21	Loss	0.986	Loss	0.995	Loss
Match 22	Win	0.732	Win	0.732	Win
Match 23	Win	0.000	Loss	0.000	Loss
Accuracy		0.751	0.826	0.720	0.783

[a]Accuracy as measured by the forecacst result at each epoch during a match.
[b]Result forecast by the model at a match's conclusion

4 Discussion

The results of this study demonstrate that accurate real-time prediction is achievable under real-world conditions using non-simulated AFL transactional data. Specifically, results showed mean epoch accuracies in excess of 71% and final forecast accuracies in excess of 78%. These epoch accuracies show that except for 4 or 5 matches the model accurately captures on-field momentum and more importantly team scores whilst final model forecasts are significantly higher than commonly accepted ex-ante prediction accuracies. When comparing our two initial probability approaches, it is important to note that they yield similar results regardless of initial probability selection. This is important for two reasons: (1) it demonstrates our models convergence to stationarity as time tends towards the end of the match and (2) it enables us to forgo the initial multinomial logistic regression model used to generate the static initial probabilities hence saving computation time and streamlining the predictive process. The scope of this research is somewhat novel and as such not much exists in literature with which to compare our results. However, our model greatly outperforms the models reported in research by Štrumbelj and Vračar [3] in which they compare their own Markov model against other common methods and sources of basketball forecasts. Lagged responses may be fixed with increased transactional granularity, and however, this may increase computational burden and model preparation times.

5 Conclusion

In conclusion, the Markov model implemented within this study has shown to be practically acceptable, obtaining greater accuracies than that of static only ex-ante models. Further research and exploration is however still needed, and as more data is made available, it is theorised that far more robust and accurate models may be created. In addition to this, more automation in terms of variable selection would also be preferable.

Acknowledgements Champion Data and AFL Tables for supplying the data used in this study.

Funding The authors would like to acknowledge the contribution of an Australian Government Research Training Program Scholarship in supporting this research.

References

1. Australian Football League: Laws of Australian football (2015)
2. K. Shirley, A markov model for basketball, in *New England Symposium for Statistics in Sports* (2007)
3. E. Štrumbelj, P. Vračar, Simulating a basketball match with a homogeneous markov model and forecasting the outcome. Int. J. Forecast. **28**(2), 532–542 (2012). https://doi.org/10.1016/j.ijforecast.2011.01.004, http://www.sciencedirect.com/science/article/pii/S0169207011000458
4. S. Akhtar, P. Scarf, Forecasting test cricket match outcomes in play. Int. J. Forecast. **28**(3), 632–643 (2012). https://doi.org/10.1016/j.ijforecast.2011.08.005, http://www.sciencedirect.com/science/article/pii/S0169207011001622
5. S. Robertson, N. Back, J.D. Bartlett, Explaining match outcome in elite australian rules football using team performance indicators. J. Sports Sci. (ahead-of-print), 1–8 (2015)
6. AFL Tables: Afl-vfl match, player and coaching stats, records and lists (2015). http://afltables.com
7. S. Reicher, The psychology of crowd dynamics, in *Blackwell Handbook of Social Psychology: Group Processes* (2001), pp. 182–208

8. R. Stefani, S. Clarke, Predictions and home advantage for australian rules football. J. Appl. Stat. **19**(2), 251–261 (1992)
9. S.R. Clarke, Home advantage in the Australian football league. J. Sports Sci. **23**(4), 375–385 (2005)
10. R Core Team: R: A Language and Environment for Statistical Computing. R Foundation for Statistical Computing, Vienna, Austria (2018). https://www.R-project.org/
11. J. Kalbfleisch, J. Lawless, The analysis of panel data under a markov assumption. J. Am. Stat. Assoc. 863–871 (1985)
12. D.W. Hosmer Jr., S. Lemeshow, R.X. Sturdivant, *Applied Logistic Regression*, vol. 398 (Wiley, New York, 2013)
13. D.O. Logofet, E.V. Lesnaya, The mathematics of markov models: what markov chains can really predict in forest successions. Ecolo. Model. **126**(2), 285 – 298 (2000). https://doi.org/10.1016/S0304-3800(00)00269-6, http://www.sciencedirect.com/science/article/pii/S0304380000002696
14. G. Marshall, R.H. Jones, Multi-state models and diabetic retinopathy. Stat. Med. **14**(18), 1975–1983 (1995)
15. C.H. Jackson, Multi-state models for panel data: the msm package for R. J. Stat. Softw. **38**(8), 1–29 (2011). http://www.jstatsoft.org/v38/i08/
16. W.N. Venables, B.D. Ripley, *Modern Applied Statistics with S* (Springer, New York, 2002)
17. V. Goulet, C. Dutang, M. Maechler, D. Firth, M. Shapira, M. Stadelmann, expm: matrix exponential, log, 'etc' (2017). https://CRAN.R-project.org/package=expm. R package version 0.999-2
18. H. Wickham, *ggplot2: Elegant Graphics for Data Analysis* (Springer, New York, 2009). http://ggplot2.org

Genetically Optimized Deep Neural Learning for Breast Cancer Prediction

Suchitra Agrawal, Aruna Tiwari, and Ishan Goel

Abstract Breast cancer is one of the most commonly occurring cancers among women. Early detection of the disease is quite tricky as it involves multiple numbers of tests to evaluate factors such as location, type and size of tumor which influence the valid identification of cancer that can be time-consuming for the doctor. To increase the efficiency of disease detection, a deep neural network with an evolutionary optimization technique is implemented which will target at searching optimal weights simultaneously for multiple neurons of the net. It involves the optimization of weights and the number of iterations required to train the machine which will help in better prediction of cancer. This is experimented to classify the breast cancer (Wisconsin Breast Cancer Dataset) into benign and malignant classes. This technique will help in reducing the time required to diagnose breast cancer at an early stage making treatment effective. The experimentation result is comparable with other state-of-the-art methods in terms of the classification accuracy.

1 Introduction

Breast cancer, although rarely diagnosed in men, is one of the major causes of cancer-related deaths in women. It adversely affects the physiological as well as psychological health of subjects and can be fatal in some cases which are apparent from its high morbidity and mortality rates. The cases of occurrence of breast cancer are ever increasing. In the USA, there was a risk of 12.8% among the women for the occurrence of breast cancer which was earlier 9.09% in the 1970s [1]. This increase

S. Agrawal (✉) · A. Tiwari · I. Goel
Department of Computer Science and Engineering, Indian Institute of Technology Indore, Indore, India
e-mail: phd1801201003@iiti.ac.in

A. Tiwari
e-mail: artiwari@iiti.ac.in

I. Goel
e-mail: cse160001023@iiti.ac.in

© Springer Nature Singapore Pte Ltd. 2020
A. K. Nagar et al. (eds.), *Soft Computing for Problem Solving 2019*,
Advances in Intelligent Systems and Computing 1139,
https://doi.org/10.1007/978-981-15-3287-0_10

in risk of breast cancer can be due to a variety of reasons including change in lifestyle, lower life expectancy or change in the environment. Irrespective of the reason, breast cancer prognosis is of huge importance in today's world. Cancer prognostic modeling is difficult since it is governed by many diverse factors. While accurate and timely prognosis may save many patients from going through painful and expensive treatments, it may also help oncologists in managing the disease more efficiently and effectively.

It has been reported in the literature that there were around 14.9 million new breast cancer cases in the year 2012. Further, this number is expected to increase to 22 million in the next 2 decades [2]. The previous study on the investigation of the consequences of late prognosis of breast cancer confirms the strong association of spreading of the disease to more advanced stages and therefore lesser probability of saving the patient's life. A study [3] concluded that women who start the treatment within three months after the development of symptoms have a much higher chance of survival than those who start the treatment after that duration.

Mammography is the primary method to detect breast cancer currently. Magnetic resonance imaging is the next alternative to mammography. Irrespective of its drawbacks, MRIs are quite popularly used for breast cancer detection. There are a few other indicators of breast cancer like architectural distortion but are not very effective. Soft computing methods can help to improve the accuracy of cancer prediction and decrease the number of false-positive and false-negative decisions. A literature review in the next section shows that there have been huge researches on the survival prediction problem but effective prognosis of breast cancer could not be found. With rapid development of genomic, proteomic and imaging technologies, this kind of molecular-scale information about patients' tumors can be readily acquired.

Accurate and reliable prognosis can provide information on the type and intensity of required therapeutic intervention. This will save many patients from going through unnecessary painful and expensive treatment protocols and at the same time alert medical practitioners in deciding on intensity of treatment for some urgent cases. The problem mainly involves classifying the cancer tumors as benign or malignant. We use neural networks combined with other computational intelligence paradigms to improve the accuracy of the model. The model recognizes patterns in the input and based on these patterns classify the tumor as benign or malignant.

The paper has been organized as follows: Sect. 2 describes some of the existing techniques; Sect. 3 details the preliminaries for the proposed work in deep learning, Sect. 4 includes the proposed work, while experimental setup including dataset description and result comparison has been explored in Sect. 5. Section 6 provides conclusion and future work.

2 Existing Work

Several studies have been reported that have focused on breast cancer survivals. These studies have applied different approaches to the given problem and achieved high classification accuracies. Details of some of the previous research works are given in the following.

Continuous advancements in the field of machine learning and soft computing have made training complex models easier especially for classification. The models proposed by [4] and [5] using deep convolutional neural network have achieved high accuracy. Also, some of the state-of-the-art models have shown great improvement in the area of image and object recognition [6]. Additionally, artificial neural networks (ANNs) are been treated as perfect substitute of traditional methods and models for computer vision, speech recognition and natural language processing [7–9]. As a matter of fact, ANNs can approximate continuous as well as arbitrary non-continuous information which predominantly is the case while dealing with cancer data.

There are several methods developed for the detection of breast cancer. Liu et al. [10] used decision table (DT)-based predictive models for breast cancer survivability, concluding that the survival rate of patients was 86.52%. They employed the under-sampling C5 technique and bagging algorithm to deal with the imbalanced problem, thus improving the predictive performance on breast cancer. Tan and Gilbert [11] demonstrated the usefulness of employing ensemble methods in classifying microarray data and presented some theoretical explanations on the performance of ensemble methods. As a result, they suggested that ensemble machine learning should be considered for the task of classifying gene expression data for cancerous samples. Chaurasia and Pal [12] compared the performance criterion of supervised learning classifiers, such as naïve Bayes, SVM-RBF kernel, RBF neural networks, decision tree (Dt) (J48) and simple classification and regression tree (CART), to find the best classifier in breast cancer datasets. The experimental result shows that SVM-RBF kernel is more accurate than other classifiers; it scores at the accuracy level of 96.84% in the Wisconsin Breast Cancer (original) Datasets.

There exist several techniques for detecting breast cancer using ANNs as well. Genetic programming (GP) is one such technique introduced by Koza et al. [13] to alter the structure and weight matrix of an ANN while Bhardwaj et al. [14] used GP to give an optimal architecture of feed-forward neural network. Least mean square error on the training set was applied to find the suitable individuals which helped to classify the breast cancer tumor as malignant or benign.

In the following section, neuro-evolution [15] approach using non-dominated sorting genetic algorithm (NSGA-III) to update the network with optimal hyper-parameters has implemented to detect the presence of cancer, along with basic deep neural network models.

3 Preliminaries

The following section describes the different models required to build the genetically optimized deep neural network.

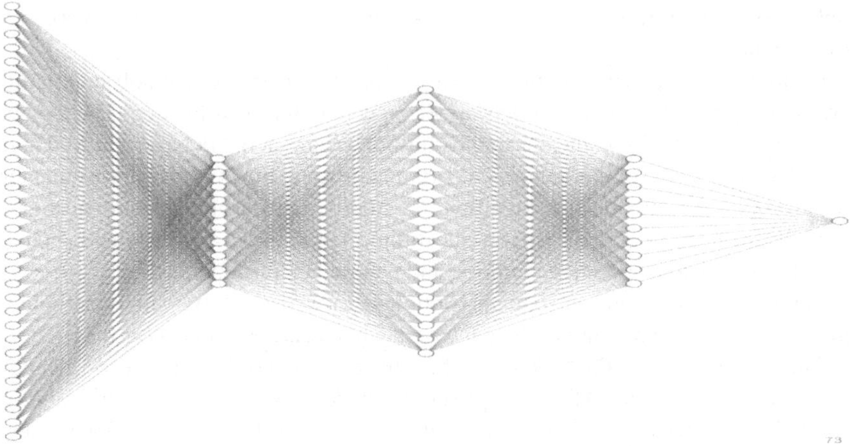

Fig. 1 Architecture overview of basic DNN

3.1 Deep Neural Network (DNN)

The deep neural networks are defined in the forms of layers in which there is an input layer denoted by $X = \{x_1, x_2, \ldots, x_n\}$, where n denotes the number of neurons in input layer, an output layer denoted by $Y = \{y_1\}$ and number of hidden layers (more than 1) in between input layer and output layer, denoted by $H1 = \{h_1, h_2, \ldots, h_i\}$, $H2 = \{h_1, h_2, \ldots, h_j\}$, $H3 = \{h_1, h_2, \ldots, h_k\}$, where i, j, k denotes the number of neurons in hidden layers $H1$, $H2$ and $H3$, respectively. The weight matrix between neuron connections in each and every layer is denoted by: $W_{ab} = \begin{pmatrix} w_{11} & \cdots & w_{1b} \\ \vdots & \ddots & \vdots \\ w_{a1} & \cdots & w_{ab} \end{pmatrix}$

The activation function used for this DNN is rectified linear unit (ReLU) which is represented as: $g = \max(0, g_{in})$, where g denotes any neuron to be processed for giving an output either from hidden layer or output layer.

The architecture implemented involves a deep neural network architecture that follows the technique of feed-forward neural network with three hidden layers having 10, 20 and 10 neurons each along with ReLU activation function. The network can be visualized as shown in Fig. 1.

3.2 Neuro-evolution Model

The second model involves preprocessing the data by introducing an encoding and an embedding layer followed by a deep neural network. Sometimes, the input data may contain some categorical variables which cannot be directly used as input for a deep neural network. There are several data encoding methods available to convert these categorical variables into continuous data. The encoding technique used here is "one-hot encoding."

One-hot encoding has two drawbacks. First, when categorical variables have high cardinality, one-hot encoding outputs large size vector which influences computational complexity. Secondly, every entity of categorical variable treated independently without considering intrinsic relations between them and this affects prognosis precision to some extent apart from over fitting data. A possible solution to these drawbacks is an appropriate transformation of categorical data into numerical vector space. This is also termed as entity embedding of the categorical data.

Binary outputs obtained from the embedding layer are then combined with the continuous variables to be fed into the deep neural network. Subsequently, a wide range of methods adopted for continuous data can be applied to this transformed numerical vector.

The implemented model can be visualized.

In the model depicted above, 5 deep layers are implemented after the embedding layer with 16 neurons in each layer using ReLU activation function.

3.3 Genetic Algorithm

Genetic algorithm is an adaptive heuristic search algorithm based on the idea of natural selection suitable for optimization problems and search problems. Evolutionary algorithms are a heuristic-based approach to solving problems that cannot be easily solved in polynomial time. The premise of genetic algorithm is with the process of natural selection. Genetic algorithm (GA) contains four overall steps: initialization, selection, genetic operators and termination. These steps each correspond, roughly, to a particular facet of natural selection and provide easy ways to modularize implementations of this algorithm category. Simply put, in a GA, fitter members will survive and proliferate, while unfit members will die off and not contribute to the gene pool of further generations, much like in natural selection. The following are the steps taken to run the genetic algorithm.

3.3.1 Selection

To propagate the best genes, we select the best of the current population. The top 50% of the population, sorted by their accuracies, are selected for mating. Sorting through loss is mathematically convenient since the precision of loss is greater than that of accuracy, but sorting by losses was found to produce a false order of the population which did not follow a strict increase in accuracy.

3.3.2 Crossover

Each individual possesses three genes which contribute to the uniqueness of individuals in the population. To produce a child, we iterate through each gene and randomly select a parent's genes to assign to the child.

3.3.3 Mutation

Every individual in the population is passed through the mutation function where each individual has a 1 in 100 chance to mutate; i.e., we assign random values to the individual's genes. This is done to enrich the gene pool and avoid problems like early convergence and stale iterations where there is no significant change in the best of the population.

4 Proposed Work

4.1 Genetically Optimized Deep Neural Network (GODNN)

The proposed model involves the amalgamation of neural network architecture and an evolutionary algorithm. This model consists of the network architecture mentioned in Sects. 3.1, 3.2 and 3.3. It uses the genetic algorithm to optimize the hyper-parameters of the network (such as the number of layers, number of nodes and activation function used). The following are the specifics of the implementation of the genetically optimized deep neural network model. The neuro-evolutionary algorithm follows the reinforcement learning paradigm as compared to conventional deep learning techniques because it restricts them from getting stuck into local minimal. Also it is important to mention that we perform all the below calculations with a fixed random seed; otherwise, the algorithm fails to converge.

The algorithm flow is implemented as shown in Fig. 3 combining it with layers mentioned in Fig. 2.

We first initialize the population to have random genes over the specified ranges (maximum number of layers/nodes, activation functions). This is done to get a good gene pool of individuals.

The population goes through the selection, crossover and mutation processes in the given order. We terminate the execution if any of the following convergence criteria are satisfied at the end of a single iteration.

- The maximum number of iterations is reached.
- The best neural net achieves a set accuracy for given dataset.

The following section describes the results achieved using above mentioned models along with the dataset description and other technical aspects.

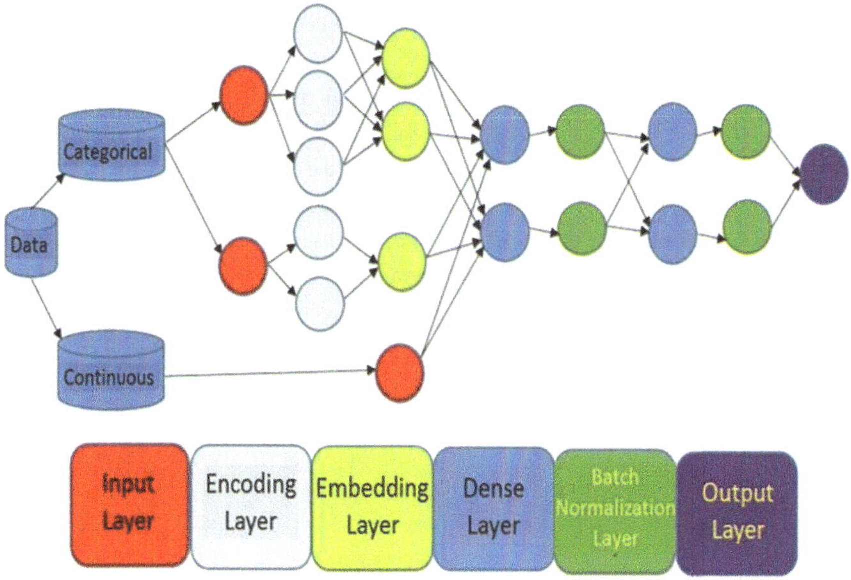

Fig. 2 Architecture overview using one-hot encoding

Fig. 3 Lifecycle of genetic algorithm

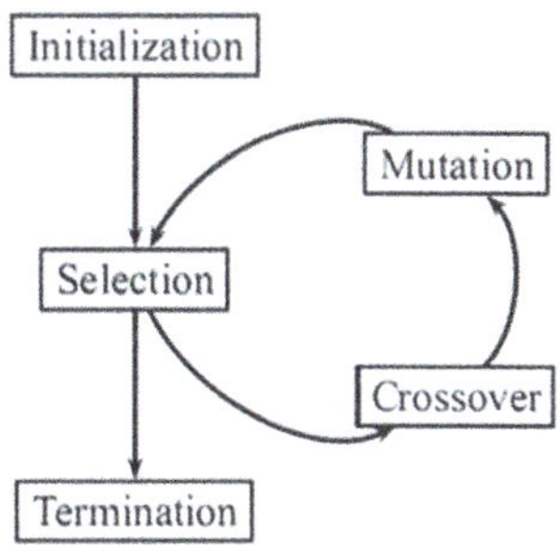

5 Experimental Results

The proposed GODNN model along with DNN and neuro-evolutionary model is experimented for prediction of breast cancer. We run the models on Jupyter Notebook platform using Keras and TensorFlow library under a desktop that equips with Intel i3 3.10 GHz CPU, 10 GB memory. The dataset details and results for each model are described next.

Table 1 UCI ML Wisconsin Breast Cancer Dataset feature description

S. no.	Feature	Description
1	Radius	Mean of distances from center to points on the perimeter
2	Texture	Standard deviation of gray-scale values
3	Perimeter	Boundary distance of the cell nucleus
4	Area	Area covered by cell nucleus
5	Smoothness	Local variation in radius lengths
6	Compactness	$\text{Perimeter}^2/\text{area} - 1.0$
7	Concavity	Severity of concave portions of the contour
8	Concave points	Number of concave portions of the contour
9	Symmetry	Symmetry of cell nucleus
10	Fractal dimension	"Coastline approximation" − 1

5.1 Dataset

The dataset used in this project is Wisconsin Breast Cancer (Diagnostic) Dataset [16]. It is a multivariate dataset with 32 attributes and 569 examples. The dataset used consists of labeled data where the attributes describe features of the tumor. The dataset also involves some categorical variables which cannot be represented as a mathematical vector directly. The features included in the dataset are mentioned in Table 1. The mean, standard error and worst or largest (mean of the three largest values) of these features were computed for each image, resulting in 30 features. Therefore, each object has 30 dimensions and the target labels are binary with values of 1 and 0. The dataset used in the following test runs is the University of California Irvine Machine Learning Repository (UCI ML) Wisconsin Breast Cancer (Diagnostic) Dataset.

The dataset has a total of 569 objects for which we used the 80/20 rule for dividing the dataset into training and testing set. The accuracies and loss in the fitness function are calculated over the test dataset.

5.2 Result

We present the results of each model one by one and a deep insight into the results of GODNN model. We trained the above architectures with 455 samples and performed testing with 114 samples, following the 80/20 rule for dividing data into training and testing set.

5.2.1 DNN Model

In this model, as mentioned in Sect. 3, there are 3 hidden layers with number of neurons as 10, 20 and 10, respectively. We run the algorithm for 2000 iterations as after 2000 iterations, we get a steady accuracy for an individual run. The accuracy obtained from this model is not consistent for multiple runs due to random initialization of weight matrix and thus can vary from 95 to 98.5% and the mean is found to be 95.5%.

5.2.2 Neuro-evolution Model

In this architecture, we run the model for 2800 iterations, which is 5 epochs, which get us a steady accuracy of 98% within 4 epochs and negligible changes occurred during the 5th epoch. The neuro-evolution model is advantageous such that it can always be used irrespective of the input data format, whereas the DNN model required the input data to be continuous. A simple deep neural network cannot deal with categorical variables; hence, neuro-evolution model proves to be better than DNN model in every aspect.

5.2.3 GODNN Model

The results of GODNN are as follows, and also it is the culmination of the best parts of the previously mentioned models. The model was executed for 5 epochs, i.e., 2800 iterations on the whole dataset. The model was also executed in two different initialization conditions, which are as follows:

Case #1: When the best neural net in the first iteration satisfies the threshold, thus resulting in solution convergence. The accuracy of all the individuals at the end of the first iteration is given in Fig. 4.

As mentioned earlier, the second convergence criterion is satisfied in the first iteration due to which the algorithm is terminated and we get a model of 100% accuracy over the test dataset for all the remaining iterations. The second iteration test accuracy is provided in Fig. 5.

The best neural network in this run has the architecture mentioned in Fig. 6.

Case #2: We run the algorithm over the same test and training datasets. Since the initial population's genes are randomized, we may achieve our best results at different points of execution in different runs of the algorithm. The accuracies of the population at the end of the first iteration are as given in Fig. 7, while the accuracies of the population at the end of the second iteration are as given in Fig. 8.

In this case, the solution converges in the second iteration by satisfying the threshold criteria. The trend in accuracy of the best individual of the population at the end of each iteration is given in Fig. 9.

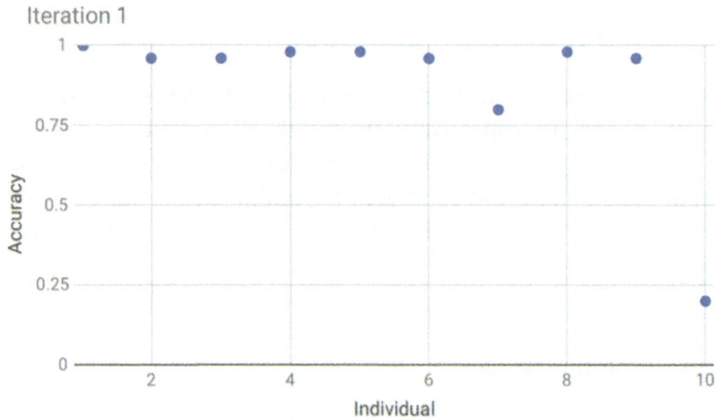

Fig. 4 Test accuracy for GODNN Case 1, Iteration 1

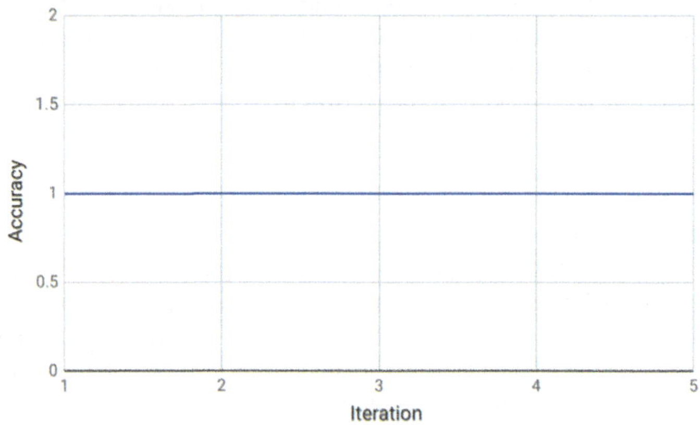

Fig. 5 Test Accuracy for GODNN Case 1, Iteration 2

Fig. 6 Architecture specifications for GODNN Case 1

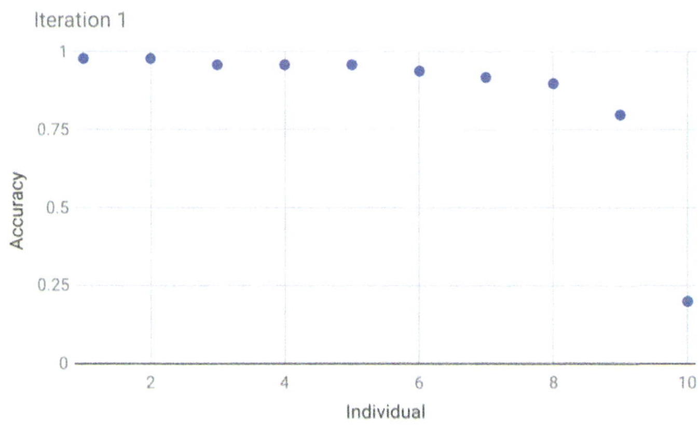

Fig. 7 Test Accuracy for GODNN Case 2, Iteration 1

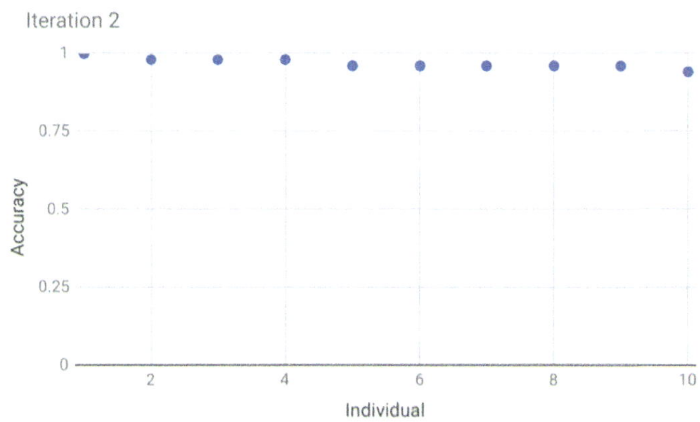

Fig. 8 Test Accuracy for GODNN Case 2, Iteration 2

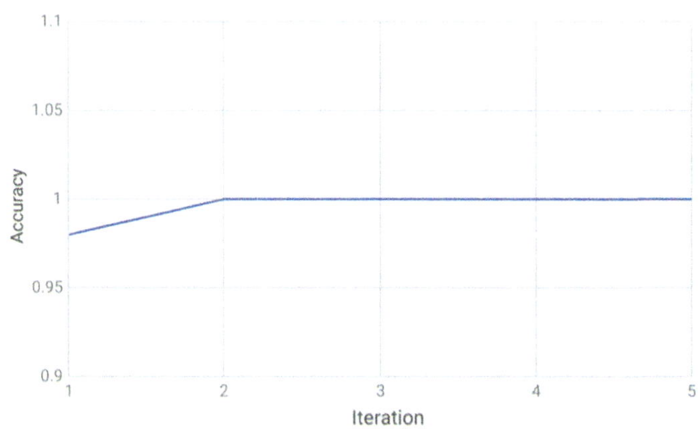

Fig. 9 Trend in accuracy of the best individual

Table 2 Test accuracy of all models on Wisconsin Dataset

Model		Testing accuracy (%)
Architecture 1		95
Architecture 2		98
Architecture 3	Case 1	100
	Case 2	100

Executing the genetic algorithm using deep neural network over the mentioned dataset, we achieve a steady accuracy of 100%. Table 2 provides the results of the models tested with Wisconsin Breast Cancer Dataset.

6 Conclusion

Summarizing the research and data stated above, there were three models implemented over the UCI ML Wisconsin Breast Cancer Dataset. Model I achieved a varying accuracy in the range of 95 and 98.5% over the test dataset. Model II achieved a steady accuracy of 98% over the test dataset. The proposed GODNN model has a variable architecture and generated according to the genetic algorithm. The accuracy achieved was consistently 100% over the test dataset. From the above results, GODNN is observed to be the best approach beating the other two models by a substantial margin in terms of accuracy. This is because a combination of neuro-evolution and entity embedding always generates a feed-forward neural network which achieves the best generalization over the dataset.

References

1. B. Abdikenov, Z. Iklassov, A. Sharipov, S. Hussain, P.K. Jamwal, Analytics of heterogeneous breast cancer data using neuroevolution. IEEE Access **7**:18,050–18,060 (2019). https://doi.org/10.1109/access.2019.2897078
2. N. Agarwal, V.N. Balasubramanian, C. Jawahar, Improving multiclass classification by deep networks using dagsvm and triplet loss. Pattern Recogn. Lett. **112**, 184–190 (2018). https://doi.org/10.1016/j.patrec.2018.06.034
3. J. Ahmad, K. Muhammad, J. Lloret, S.W. Baik, Efficient conversion of deep features to compact binary codes using fourier decomposition for multimedia big data. IEEE Trans. Industr. Inf. **14**(7), 3205–3215 (2018). https://doi.org/10.1109/TII.2018.2800163
4. J. Ferlay, I. Soerjomataram, R. Dikshit, S. Eser, C. Mathers, M. Rebelo, D. Parkin, D. Forman, F. Bray, Cancer incidence and mortality worldwide: sources, methods and major patterns in globocan 2012. Int. J. Cancer **136**, E359–E386 (2015)
5. F. Gao, T. Wu, J. Li, B. Zheng, L. Ruan, D. Shang, B. Patel, Sd-cnn: a shallow deep cnn for improved breast cancer diagnosis. Comput. Med. Imag. Graph. **70**, 53–62. https://doi.org/10.1016/j.compmedimag.2018.09.004, http://www.sciencedirect.com/science/article/pii/S0895611118302349

6. J. Gu, Z. Wang, J. Kuen, L. Ma, A. Shahroudy, B. Shuai, T. Liu, X. Wang, G. Wang, J. Cai, T. Chen, Recent advances in convolutional neural networks. Pattern Recogn. **77**, 354–377 (2018). https://doi.org/10.1016/j.patcog.2017.10.013, http://www.sciencedirect.com/science/article/pii/S0031320317304120
7. E. Hortal, D. Planelles, F. Resquin, J.M. Climent, J.M. Azor, J.L. Pons, Using a brain-machine interface to control a hybrid upper limb exoskeleton during rehabilitation of patients with neurological conditions. J. Neuro Eng. Rehabilit. **12**(1), 92 (2015). https://doi.org/10.1186/s12984-015-0082-9
8. J. Liu, B. Xu, C. Zheng, Y. Gong, J. Garibaldi, D. Soria, A. Green, I.O. Ellis, W. Zou, G. Qiu, An end-to-end deep learning histochemical scoring system for breast cancer tma. IEEE Trans. Med. Imaging **38**(2), 617–628 (2019). https://doi.org/10.1109/TMI.2018.2868333
9. Y. Liu, C. Wang, L. Zhang, Decision tree based predictive models for breast cancer survivability on imbalanced data, in *2009 3rd International Conference on Bioinformatics and Biomedical Engineering*, pp. 1–4 (2009). https://doi.org/10.1109/icbbe.2009.5162571
10. S. Purushotham, C. Meng, Z. Che, Y. Liu, Benchmarking deep learning models on large healthcare datasets. J. Biomed. Inform. **83**, 112–134 (2018). https://doi.org/10.1016/j.jbi.2018.04.007
11. M. Richards, A. Westcombe, S. Love, P. Littlejohns, A. Ramirez, Influence of delay on survival in patients with breast cancer: a systematic review. The Lancet **353**(9159), 10–241,119–1126. https://doi.org/10.1016/s0140-6736(99)02143-1
12. A.C. Tan, D. Gilbert, Ensemble machine learning on gene expression data for cancer classification. Appl. Bioinformat. **2**, S75–S83 (2003)
13. J.R. Koza, J.P. Rice, Genetic generation of both the weights and architecture for a neural network, in *IJCNN-91-Seattle International Joint Conference on Neural Networks*, vol. ii, pp. 397–404. https://doi.org/10.1109/ijcnn.1991.155366
14. A. Bhardwaj, A. Tiwari, Breast cancer diagnosis using genetically optimized neural network model. Expert Syst. Appl. **42**(10), 4611–4620 (2015). https://doi.org/10.1016/j.eswa.2015.01.065
15. V. Chaurasi, S. Pal, Data mining techniques: to predict and resolve breast cancer survivability. Int. J. Comput. Sci. Mobile Comput. **3**, 10–22
16. UCI ML Breast Cancer Wisconsin (Diagnostic) dataset. https://archive.ics.uci.edu/ml/datasets/Breast+Cancer+Wisconsin+(Diagnostic)

Optimization of Lycopene Extraction from Tomato Processing Waste Skin Using Harmony Search Algorithm

Assif Assad, Kusum Deep, Neil Buckley, and Atulya K. Nagar

Abstract The skin of a tomato (Lycopersicon esculentum) is a rich source of lycopene, an important component of waste from tomato paste manufacturing plants. The lycopene content in the skin of a tomato pomace is about five times higher than in the pulp. According to the World Processing Tomato Council, 1,200,000 tons of tomato processing waste is produced worldwide annually. In this study, lycopene extraction from tomato skins is formulated as an optimization problem with respect to five independent variables, namely solvent/meal ratio, temperature, number of extractions, extraction time and particle size. The music-inspired harmony search (HS) metaheuristic algorithm is used to optimize the lycopene production, and the maximum lycopene (4.8 mg/100 g) was predicted when the solvent/meal ratio, temperature, number of extractions, extraction time and particle size is 20:1 v/w, 60 C, 5, 4 min and 0.43 mm, respectively. Response surface analysis (RSM) has been used in (Kaur et al, Food Chem 2008 [1]) to optimize the lycopene yield with respect to these variables, and the maximum yield predicted was 1.99 mg/100 g. Thus, the methodology applied in the present work for optimizing the extraction of lycopene is far better than response surface analysis. Further comparison with particle swarm optimization reveals that HS is more reliable optimization technique to solve the problem at hand.

Keywords Lycopene · Harmony search · Metaheuristics · Evolutionary algorithms

---

A. Assad (✉)
Department of Computer Science and Engineering, Islamic University of Science and Technology, Awantipora 192122, India
e-mail: assifassad@gmail.com

K. Deep
Department of Mathematics, Indian Institute of Technology Roorkee, Roorkee 247667, India

N. Buckley · A. K. Nagar
Department of Mathematics and Computer Science, Faculty of Science, Liverpool Hope University, Liverpool L16 9JD, UK

© Springer Nature Singapore Pte Ltd. 2020

A. K. Nagar et al. (eds.), *Soft Computing for Problem Solving 2019*,
Advances in Intelligent Systems and Computing 1139,
https://doi.org/10.1007/978-981-15-3287-0_11

1 Introduction

Lycopene belongs to the carotenoid family and has a bright red pigment that has been studied extensively due to its interesting biological activities. It is a potent antioxidant, protecting cells against oxidative damage and reducing the risk of many chronic diseases [2]. Various studies have shown that lycopene is effective in decreasing the risk of prostate and breast cancer [3, 4]. Some studies also revealed its effectiveness in the prevention of coronary and cardiovascular diseases. Indeed, it inhibits the activation of inducible nitric oxide syntheses proteins, and hence it has anti-inflammatory properties [5]. It reduces blood cholesterol levels by preventing low-density lipoprotein oxidation.

In the food industry, it is used as a food additive to enhance nutritional properties and storage stability [6]. According to the World Processing Tomato Council, 1,200,000 tons of tomato processing waste is produced worldwide annually [7]. One of the main components of this waste is skin. The lycopene content present in the skin of tomato pomace is approximately five times higher than in the pulp [8]. Lycopene is found at a concentration of 4–14.3 mg/100 g in common variants of tomato [9]. The processing waste is either used as animal fodder or discarded, but its abundance in lycopene makes it a promising prospect as a sustainable, low-cost alternative of this nutraceutical compound.

Lycopene from tomatoes is usually extracted using organic solvents such as hexane, ethanol, acetone and chloroform. In addition, some recent studies have shown that lycopene could also be extracted using various advanced methods such as ultrasound-assisted extraction, ultrasound/microwave-assisted extraction, supercritical fluid extraction, enzyme-assisted extraction and selective inclusion in deoxycholic acid [10].

Kaur et al. studied the effect of five independent values, namely solvent/meal ratio, temperature, number of extractions, extraction time and particle size on lycopene extraction [1]. The optimal yield of lycopene predicted using response surface analysis is 1.97 mg/100 g and the optimal yield experimentally obtained is 1.99 mg/100 g [1].

Poojary and Passamonti studied the effect of using acetone/n-hexane mixtures at different ratios (1:3, 2:2 and 3:1, v/v) and at different temperatures (30, 40 and 50 °C) on lycopene extraction and the yield obtained was in the range 3.47–4.03 mg/100 g [11].

The extraction of pure all-trans-lycopene from the pulp fractions of tomato processing waste was studied in [10]. A full factorial design consisting of four independent variables including extraction temperature, time, percentage of acetone in n-hexane and solvent volume was used to investigate the effects of process variables on the extraction. The absolute amount of lycopene present in the pulp waste was found out to be 3.8 mg/100 g using response surface analysis.

Ultrasound-assisted extraction of lycopene from tomato processing wastes was studied in [12]. It was showed that the ultrasound-assisted extraction of lycopene requires less time, lower temperature and lower solvent when compared with conventional organic solvent extraction.

Response surface methodology was employed to optimize lycopene extraction with solvent ratio solid–liquid ratios, microwave power and delivered energy equivalents as factors in [13]. At optimal setting, the all-trans-lycopene yield obtained was 13.592 mg/100 g.

Lycopene extraction was carried out via the ultrasonic-assisted extraction and the process was optimized with response surface methodology in [14]. Sonication enhanced the efficiency of relative lycopene yield, lowered the extraction temperature and shortened the total extraction time.

High hydrostatic pressure-assisted extraction was utilized for lycopene extraction from tomato pulp in [15]. Employing RSM, three parameters were optimized for lycopene extraction and the optimal yield of lycopene obtained was 2.01 mg/100 g.

The rest of the paper is organized as follows: Sect. 2 introduces the harmony search algorithm, Sect. 3 describes the materials and methods used, Sect. 4 describes experimentation and the paper is concluded in Sect. 5.

2 Harmony Search Algorithm (HS)

Optimization is defined as the process of selecting the best option from a set of available alternatives. Every process has the potential to be optimized and many challenging problems in science, business, economics and engineering can be formulated as an optimization process.

Many real-life optimization problems are complex and thus difficult to solve in an exact manner within reasonable amount of time. The classical optimization methods have the limitation of being highly sensitive to the starting point and frequently converge to a local optimum. Classical optimization techniques can be broadly classified into two categories: direct search method and gradient-based search method. In the direct search method, only the objective function and constraints are utilized for the search process, whereas in the gradient-based search method, the first-order and/or second-order derivatives are utilized in the search process. Direct search methods have the limitation of slow convergence, whereas gradient-based search methods are faster, however, have the limitation of being non-applicable to discontinuous and non-differentiable functions. Furthermore, both methods seek local optima, thus starting the search in the vicinity of a local optima causes them to miss the global optima.

Metaheuristic algorithms eradicate some of the afore-mentioned difficulties and are quickly replacing the classical methods in solving complex nonlinear optimization problems. Metaheuristic algorithms typically intend to find a reasonably good solution close to optimization in reasonable amount of computational time. During the last few decades, several metaheuristic algorithms have been proposed including genetic algorithms (GA), genetic programming, evolutionary strategies (ES), evolutionary programming, differential evolution (DE), ant colony optimization (ACO) and particle swarm optimization (PSO) to name a few.

In 2011, Geem et al. created a new evolutionary algorithm called harmony search, inspired by optimal musical composition [16]. HS has been used to solve diverse problems in science and engineering, including structural design [17], multi-objective optimization [18], load dispatch problems in electrical engineering [19] and classification and feature selection [20, 21]. A detailed literature review on recent applications of HS can be found in [22, 23].

In order to explain the workings of HS, let us idealize the improvisation process by a musician. When a skilled musician is improvising, he has three possible choices:

1. Play any piece of music exactly from his memory.
2. Play something similar to a known piece.
3. Compose new or random notes.

These three options were formalized by Geem et al. [16] into an optimization process, and the three corresponding components are referred to as usage of harmony memory, pitch adjusting and randomization. The usage of harmony memory (HM) is similar to the choice of the best fit individuals in genetic algorithms. A parameter called harmony memory considering rate (HMCR 2 [0, 1]) is typically used to tune HM. If HMCR is near 0 (low), the algorithm shows slow convergence because best harmonies are not utilized fully, whereas if HMCR is near 1 (very high), the solution space is not explored properly, resulting in sub-optimal solutions. The recommended range of HMCR is (HMCR 2 [0.7, 0.95]). The second parameter of HS is pitch adjustment determined by pitch adjusting rate (PAR) and fret width/bandwidth (BW). PAR corresponds to generating a slightly different solution than the solution already stored in HM. Pitch can be adjusted linearly or nonlinearly, however, most often the following linear adjustment is used.

$$H_i^{new} = H_i^{old} + BW \times r_i \quad where\, r_i \in [-1, 1] \quad and\, 1 \le i \le D \tag{1}$$

where H_i^{old} is the ith component of the existing harmony or solution and H_i^{new} is the ith component of the new harmony after the pitch adjusting action, and BW is the bandwidth. The Eq. (1) produces a new solution/harmony around the existing solution/harmony by altering it by a small random amount. Here, D is the total number of components in the harmony and r_i is a random number generated in the range of $[-1, 1]$. The recommended value of $PAR \in [0.1, 0.5]$. The third HS parameters are randomization, promoting exploration of the search space. The pseudo-code of HS is shown as Algorithm 1. H represents a potential solution (harmony), $rand \in [0, 1]$ is a uniformly distributed random number, $rand\_int(1, HMS)$ generates a uniformly distributed integer random number between 1 and HMS, size of harmony memory, is represented as HMS, and D is the dimensionality of the problem.

3 Materials and Methods

The method for extracting lycopene in this manuscript is exactly the same as defined in [1] and is briefly introduced in this section.

3.1 Material

The material for experimentation (tomato pomace) was obtained from a local tomato paste manufacturing unit located in Haridwar, Uttarakhand (India).

Algorithm 1 HARMONY SEARCH (HS) ALGORITHM

```
Define Objective function f(H), HMCR, PAR, BW and HMS.
Initialize the Harmony Memory (HM).
while (Stopping Criteria Not Reached) do
  Find current Worst and Best harmony in HM.
  for i = 1 to D do
    if (rand ≤ HMCR) then
      H_i = HM_i^j where j=rand_int(1,HMS)
      if (rand ≤ PAR) then
        H_i = H_i ± rand × BW
      end if
    else
      Generate H_i randomly within the allowed bounds.
    end if
  end for
  if (H is better than worst Harmony in HM) then
    Update HM by replacing WORST harmony by H.
  end if
end while
print Best Harmony as obtained solution.
```

3.2 Sample Preparation

A continuous floatation-cum-sedimentation system was used to separate skin from pomace. A cabin dryer was used to dry separated tomato skin according to the method described in [9]. The dried skin was then passed through different sieves of sizes 0.05, 0.15, 0.25, 0.35 and 0.43 mm.

3.3 Proximate Analysis

Ash, moisture, crude fiber, crude fat and crude protein were determined by the procedure used in [24].

3.4 Pigment Extraction

A sample of one gram was extracted using solvent (hexane:acetone:alcohol 2:1:1) containing 0.05% (w/v) butylated hydroxytoluene (BHT). 15 ml of cold distilled water was added to the sample, followed by the agitation of suspension. In order to separate polar and non-polar layers (Fig. 1), the solution was allowed to stand for 15 min. The layer containing lycopene (polar layer) was extracted. A UV visible spectrophotometer (Fig. 2) was then used to measure absorbance at 471 nm and expressed as mg/100 g using an extinction coefficient of 17:2 104 mol cm^{-1}.

4 Experimental Design

In order to formulate lycopene extraction as an optimization problem w.r.t five independent variables, X1 (solvent/meal ratio), X2 (number of extractions), X3 (temperature), X4 (particle size) and X5 (time), and 32 experiments were carried out. Table 2 shows the values of these variables along with the content of lycopene obtained

Fig. 1 Polar and non-polar layer

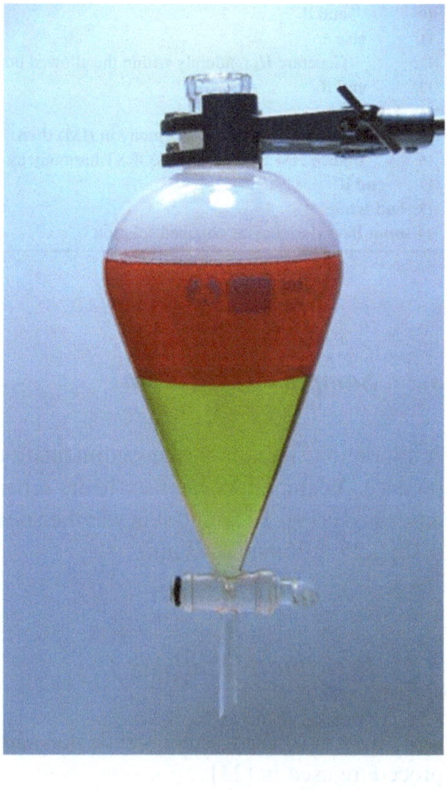

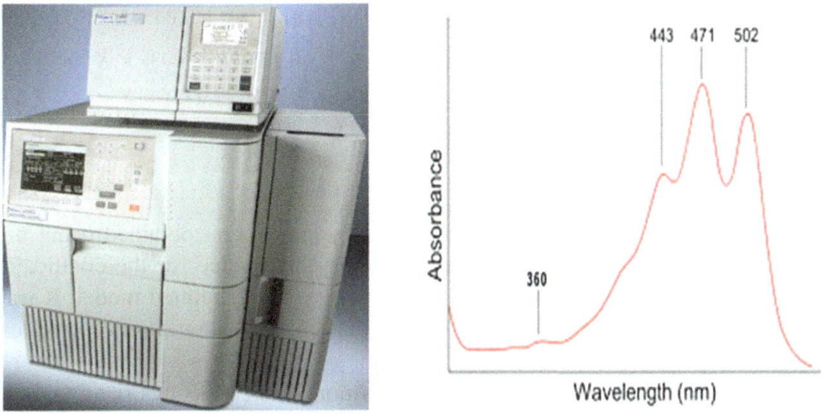

(a) HPLC-UV Visible Spectrometer (b) Absorption spectrum of lycopene in hexane

Fig. 2 HPLC system and absorption spectrum of lycopene

experimentally. Table 1 shows the independent variables and their levels used for central composite design.

Least square curve fitting was used to obtain a quadratic regression equation of the form of Eq. 2, depicting the relation between the five independent variables and the dependent variable (lycopene yield), the regression equation is given in Eq. 3.

$$Y = b_0 + \sum_{n=0}^{5} b_n X_n + \sum_{n=0}^{5} b_{mn} X_n^2 + \sum_{n<m}^{5} b_{mn} X_n X_m \tag{2}$$

where Y is the lycopene yield (mg/g), b_0 is the value for the fixed response at the central point of the experiment. b_n, b_m, b_{nm} are, respectively, the linear, quadratic and cross-product coefficients.

$$Y = 0.86306818181 - 0.03750 \times X_1 + 0.10791666666667 \times X_2 + 0.077750$$
$$\times X_3 + 0.070250 \times X_4 + 0.0638333 \times X_5 + 0.11868181 \times X_1^2 + 0.1143068181$$

Table 1 Independent variables and their levels used for central composite design

Independent variables	Symbol	Coded variable levels				
		−2	−1	0	1	2
Solvent/meal ratio (v/w)	X_1	20	30	40	50	60
Number of extractions	X_2	1	2	3	4	5
Temperature (°C)	X_3	20	30	40	50	60
Particle size (mm)	X_4	0.05	0.15	0.25	0.35	0.43
Time (min)	X_5	4	8	12	16	20

$$\times X_2^2 + 0.14555681818182 \times X_3^2 - 0.0124431818181 \times X_4^2 + 0.04543181818182$$
$$\times X_5^2 - 0.11750 \times X_1 \times X_2 - 0.06650 \times X_1 \times X_3 - (3.01E - 003) \times X_1 \times X_4$$
$$+ 0.019250 \times X_1 \times X_5 + 0.101250 \times X_2 \times X_3 + (2.504E - 003) \times X_2$$
$$\times X_4 + 0.0150 \times X_2 \times X_5 + 0.07650 \times X_3 \times X_4$$
$$- 0.0780 \times X_3 \times X_5 - 0.020250 \times X_4 \times X_5 \quad (3)$$

The experimental and predicted value of lycopene content obtained using the regression model was compared (Table 2). A very high value for the coefficient of determination ($R2 = 0.99$) indicates the adequacy of the applied model. It was determined from statistical analysis of the data that linear, quadratic and interaction coefficients are significant.

In this study, we used the music-inspired harmony search metaheuristic to optimize (maximize) the regression equation (Eq. 3), depicting the relationship between the independent variables and the lycopene yield. The reasons for using HS are simplicity, small number of control parameters and the fact that many studies have shown it to give better results than established metaheuristics such as genetic algorithms and particle swarm optimization [25, 26].

The optimal lycopene yield predicted is 4.84 mg/100 g at parameter setting 20:1 v/w solvent/meal ratio, five extractions, 60 C temperature, 0.43 mm particle size and 4-min extraction time. In order to verify the claim, an experiential yield of lycopene obtained at this parameter setting was 4.75 mg/100 g.

In the next section, an introduction to HS is provided along with the detailed description of how it has been applied to this problem.

4.1 Proposed Methodology of Optimizing Lycopene Extraction Using Harmony Search Algorithm

The value of HMCR and PAR adopted for HS (Algorithm 1) in this experimentation are, respectively, set to 0.9 and 0.3, however, it was observed that small values of HMS produce inferior results occasionally, so HMS was set to 100. The objective function to be maximized is given in Eq. 3. Each harmony is a five-dimensional vector such that the first dimension corresponds to independent variable X1 (solvent/meal ratio, v/w), the second dimension corresponds to X2 (number of extractions), the third dimension corresponds to X3 (temperature, °C), the fourth dimension corresponds to X4 (particle size, mm) and the fifth dimension corresponds to X5 (time, min). The range of all five independent variables is from −2 to 2 in coded form as given in Table 1.

The harmony memory is initialized such that each component of the harmony (solution) is within the bounds −2 to 2. Note that the third component of harmony corresponding to X_3 (no. of extractions) can take only integer values. Each component

Table 2 Central composite arrangement for independent variables X_1 (solvent/meal ratio, v/w), X_2 (number of extractions), X_3 (temperature, C), X_4 (particle size, mm) and X_5 (time, min) and their response (lycopene yield, mg/100 g)

Run	Variables levels (encoded)					Lycopene yield (mg/100 g)	
	X_1	X_2	X_3	X_4	X_5	Experimental ± Std	Predicted
1	1	1	−1	−1	1	1.312 0:032	1.3181
2	1	1	1	−1	−1	1.168 0:027	1.1534
3	−1	1	−1	1	1	1.39 0:030	1.4086
4	1	−1	1	1	−1	1.348 0:0252	1.3281
5	−2	0	0	0	0	1.402 0:031	1.4128
6	0	2	0	0	0	1.557 0:041	1.5361
7	−1	−1	−1	−1	1	1.187 0:026	1.1773
8	−1	−1	1	−1	−1	1.041 0:024	1.0106
9	1	−1	1	−1	1	1.058 0:041	1.0258
10	−1	−1	−1	1	−1	0.948 0:043	0.9506
11	1	−1	−1	1	1	1.45 0:021	1.4508
12	−1	1	1	1	−1	1.976 0:027	1.9739
13	1	1	1	1	1	1.49 0:032	1.4861
14	0	0	0	0	0	0.846 0:036	0.8631
15	−1	1	−1	−1	−1	1.127 0:022	1.1349
16	1	1	−1	1	−1	0.934 0:033	0.9524
17	−1	−1	1	1	1	1.228 0:035	1.2083
18	−1	1	1	−1	1	1.647 0:052	1.6326
19	0	0	−2	0	0	1.32 0:043	1.2898
20	0	0	2	0	0	1.545 0:027	1.6008
21	2	0	0	0	0	1.248 0:022	1.2628
22	0	0	0	0	0	0.893 0:038	0.8631
23	1	−1	−1	−1	−1	1.192 0:031	1.1821
24	0	0	0	0	2	1.158 0:029	1.1725
25	0	0	0	−2	0	0.637 0:027	0.6728
26	0	0	0	2	0	0.964 0:034	0.9538
27	0	0	0	0	0	0.923 0:034	0.8631
28	0	−2	0	0	0	1.058 0:029	1.1045
29	0	0	0	0	0	0.832 0:032	0.8631
30	0	0	0	0	0	0.863 0:033	0.8631
31	0	0	0	0	−2	0.906 0:036	0.9171
32	0	0	0	0	0	0.847 0:021	0.8631

of the new harmony represented as H_i is generated either using HM or randomization, and it is slightly altered in step 12 (Algorithm 1) determined by the parameter bandwidth (BW). The value of BW (except X_3) for each component of harmony is 0.04 (i.e., $\frac{1}{100}$th of variable bounds) and for X_3, it is 1. Once the new harmony (H) is generated, it is evaluated using the objective function and replaces the worst harmony in HM in the case it is better than the worst current harmony. The procedure is repeated for 5000 generations.

The optimal yield of lycopene predicted using HS is 4.84 mg/100 g at parameter setting 20:1 (v/w solvent/meal ratio), five extractions, 60 °C temperature, 0.43 mm particle size and 4 min extraction time. In the coded form, the optimal values of independent variables $X1$, $X2$, $X3$, $X4$ and $X5$ obtained using HS are, respectively, −2, 2, 2, 2 and −2; hence, at these values, the objective function (Eq. 3) evaluates to 4.8. The program was run for 50 independent runs and the yield predicted was exactly the same (4.8 mg/100 g) in each run, thus the standard deviation is zero. The program was coded in Dev C++ 5.0 and was run on a laptop with Windows 10 operating system, Intel Core i3 processor and 4 GB of RAM. The execution time was approximately 0.18 s.

Figure 3 graphically shows lycopene yield predicted by HS with respect to function evaluations. Here, the initial predicted lycopene yield is about 1.6 mg/100 g. The algorithm makes steep progress in the first quarter of execution and converges near the end.

The optimal parameter setting obtained by HS is 20:1 (v/w solvent/meal ratio), five extractions, 60 °C temperature, 0.43 mm particle size and 4 min extraction time. In order to verify the claim, the experimentation was carried out at this parameter setting, the experiment was repeated three times and the lycopene yield obtained is shown in Table 3. The average yield of lycopene obtained at the optimal parameter setting is 4.75 mg/100 g experiential. Note that the experimental procedure adopted for lycopene extraction in this study is exactly same as given in [1] and has been described in Sect. 3.

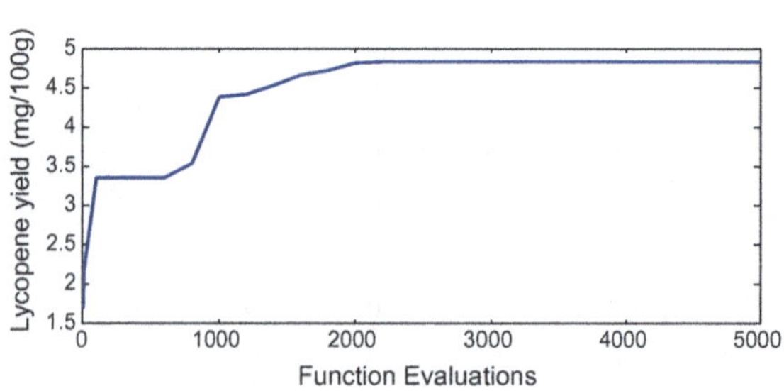

Fig. 3 Predicted lycopene yield w.r.t. no. of function evaluations

Table 3 Lycopene yield (mg/100 g) obtained at optimal setting

Run	Variables levels (un-coded)					Lycopene yield (mg/100 g)
	X_1	X_2	X_3	X_4	X_5	Experimental
1	−2(20)	2(5)	2(60)	2(0.43)	−2(4)	4.78
2	−2(20)	2(5)	2(60)	2(0.43)	−2(4)	4.72
3	−2(20)	2(5)	2(60)	2(0.43)	−2(4)	4.76
Average (std)						**4.75** (0.0305)

4.2 Comparison of Harmony Search with Particle Swarm Optimization

Particle swarm optimization (PSO), a nature-inspired optimization technique, has attracted significant attention since introduced by Kennedy and Eberhart in [27]. Simulating the social behavior of birds flocking or fish schooling, a population of particles in a PSO algorithm cooperates and interacts to search for solutions in the problem space. Owing to its conceptual simplicity and high efficiency, PSO has been successful in solving a variety of problems in many areas such as feature selection [28], intrusion detection [29], microwave imaging [30], electromagnetic inverse problems [31] and high-speed backplane connector [32]. Further HS algorithm was compared with PSO to solve the optimization problem at hand.

The particle swarm optimization algorithm contains the collection of particles that move around the search space influenced by their own best past location and the best past location of the whole swarm. In each iteration a particle's velocity and position are updated using Eqs. 4 and 5, respectively.

$$V_i(t+1) = V_i(t) + C_1 r_1 \left(p_i^{best} - p_i^t\right) + C_2 r_2 \left(p_{gbest} - p_i^t\right) \tag{4}$$

$$P_i(t+1) = P_i(t) + V_i(t) \tag{5}$$

where $V_i(t+1)$ is the updated velocity for the ith particle at time $t+1$, C_1 and C_2 are the weighting coefficients for personal best and global best positions, respectively, $P_i(t)$ is the ith particles position at time t, p_i^{best} is the ith particle's best-known position and p_{gbest} is the best position known to the swarm. The r_1 and r_2 are uniformly distributed random variable $\in [0, 1]$.

In this manuscript, the efficiency of PSO algorithm is compared with HS algorithm in solving the problem at hand. The performance of the metaheuristic algorithm depends on the tuning of hyperparameter and the parameter setting adopted for PSO in this work is same as given in [33]: C_1, C_2 are both 2 and population size is 50. In order to keep the comparison fair, both algorithms are run for equal number of function evaluations.

Each particle is a five-dimensional vector such that the first dimension corresponds to independent variable X_1 (solvent/meal ratio, v/w), the second dimension corresponds to X_2 (number of extractions), the third dimension corresponds to X_3 (temperature, °C), the fourth dimension corresponds to X_4 (particle size, mm) and the fifth dimension corresponds to X_5 (time, min). The range of all five independent variables is from −2 to 2 in coded form as given in Table 1. The swarm is initialized such that each component of the particle (solution) is within the bounds −2 to 2. Note that the third component of solution corresponding to $X3$ (no. of extractions) can take only integer values, thus it is always rounded off to the nearest integer within the bounds. The procedure is repeated for 100 generations (i.e., 5000 function evaluations).

The program was coded in Dev C++ 5.0 and was run on a laptop with Windows 10 operating system, Intel Core i3 processor and 4 GB of RAM. The execution time was approximately 0.163 s. The program was repeated for 50 independent runs.

Even though the optimal yield of lycopene predicted using PSO in 50 runs is exactly the same as predicted by HS, i.e., 4.84 mg/100 g at parameter setting 20:1 (v/w solvent/meal ratio), five extractions, 60 °C temperature, 0.43 mm particle size and 4 min extraction time. However, in few runs, the PSO got stuck in local optimal and was not able to escape. The average of lycopene predicted by PSO in 50 runs is 4.12 mg/100 g with standard deviation of ±0:6. Thus, it can be concluded that HS is more reliable to optimize the problem of lycopene extraction as compared to PSO, however, PSO shows slight efficiency in terms of time complexity.

5 Conclusion and Claim

In the first part of this work, the quadratic regression equation depicting the effect of five independent variables (solvent/meal ratio, number of extractions, temperature, particle size and time) on lycopene yield was formulated using least square fitting.

In the second stage, the regression equation was optimized with the harmony search metaheuristic algorithm. The optimal yield of lycopene predicted by HS is 4.84 mg/100 g when the parameters were set to 20:1 (v/w solvent/meal ratio), five extractions, 60 °C temperature, 0.43 mm particle size and 4 min extraction time.

In order to verify the claim, experiments were carried out at the above-mentioned optimal parameter setting obtained by HS. This showed that the lycopene yield obtained was 4.75 mg/100 g. Thus, the optimal setting predicted by HS is in tune with the experimental results.

References

1. D. Kaur, A.A. Wani, D. Oberoi, D. Sogi, Effect of extraction conditions on lycopene extractions from tomato processing waste skin using response surface methodology. Food Chem. **108**(2), 711–718 (2008)
2. A. Rao, S. Agarwal, Role of lycopene as antioxidant carotenoid in the prevention of chronic diseases: a review. Nutr. Res. **19**(2), 305–323 (1999)
3. N. Chalabi, L. Delort, L. Le Corre, S. Satih, Y.-J. Bignon, D. Bernard-Gallon, *Gene signature of breast cancer cell lines treated with lycopene* (2006)
4. N.P. Holzapfel, B. M. Holzapfel, S. Champ, J. Feldthusen, J. Clements, D.W. Hutmacher, The potential role of lycopene for the prevention and therapy of prostate cancer: from molecular mechanisms to clinical evidence. Int. J. Mol. Sci. **14**(7), 14 620–14 646 (2013)
5. M.M. Rafi, P.N. Yadav, M. Reyes, Lycopene inhibits lps-induced proinflammatory mediator inducible nitric oxide synthase in mouse macrophage cells. J. Food Sci. **72**(1), S069–S074 (2007)
6. M. Østerlie, J. Lerfall, Lycopene from tomato products added minced meat: effect on storage quality and colour. Food Res. Int. **38**(8), 925–929 (2005)
7. A. Zuorro, M. Fidaleo, R. Lavecchia, Enzyme-assisted extraction of lycopene from tomato processing waste. Enzyme Microbial Technol. **49**(6), 567–573 (2011)
8. E.H. Papaioannou, A.J. Karabelas, Lycopene recovery from tomato peel under mild conditions assisted by enzymatic pre-treatment and non-ionic surfactants. Acta Biochim. Pol. **59**(1), 71 (2012)
9. D. Kaur, R. Sharma, A. Abas Wani, B. Singh Gill, D. Sogi, Physicochemical changes in seven tomato (lycopersicon esculentum) cultivars during ripening. Int. J. Food Prop. **9**(4), 747–757 (2006)
10. M.M. Poojary, P. Passamonti, Optimization of extraction of high purity all-trans-lycopene from tomato pulp waste. Food Chem. **188**, 84–91 (2015)
11. Mahesha M. Poojary, Paolo Passamonti, Extraction of lycopene from tomato processing waste: kinetics and modelling. Food Chem. **173**, 943–950 (2015)
12. S. Kumcuoglu, T. Yilmaz, S. Tavman, Ultrasound assisted extraction of lycopene from tomato processing wastes. J. Food Sci. Technol. **51**(12), 4102–4107 (2014)
13. K. Ho, M. Ferruzzi, A. Liceaga, M. San Mart´ın-Gonzalez, Microwave-assisted extraction of lycopene in tomato peels: effect of extraction conditions on all-trans and cis-isomer yields. LWT-Food Sci. Technol. **62**(1), 160–168 (2015)
14. A.L.-S. Eh, S.-G. Teoh, Novel modified ultrasonication technique for the extraction of lycopene from tomatoes. Ultrason. Sonochem. **19**(1), 151–159 (2012)
15. V. Briones-Labarca, C. Giovagnoli-Vicuna, R. Ca ˜ nas-Saraz ˜ ua, Optimization of extraction yield flavonoids and lycopene from tomato pulp by high hydrostatic pressure-assisted extraction. Food Chem. **278**, 751–759 (2019)
16. Z.W. Geem, J.H. Kim, G. Loganathan, A new heuristic optimization algorithm: harmony search. Simulation **76**(2), 60–68 (2001)
17. S. Gholizadeh, A. Barzegar, Shape optimization of structures for frequency constraints by sequential harmony search algorithm. Eng. Optimizat. **45**(6), 627–646 (2013)
18. K. Nekooei, M.M. Farsangi, H. Nezamabadi-Pour, K.Y. Lee, An improved multi-objective harmony search for optimal placement of dgs in distribution systems. IEEE Trans Smart Grid **4**(1), 557–567 (2013)
19. L. Wang, L.-P. Li, An effective differential harmony search algorithm for the solving non-convex economic load dispatch problems. Int. J. Electr. Power Energy Syst. **44**(1), 832–843 (2013)
20. R. Diao, Q. Shen, Feature selection with harmony search. IEEE Trans. Syst. Man Cybernet. Part B Cybernet. **42**(6), 1509–1523 (2012)
21. H. Fattahi, A. Gholami, M.S. Amiribakhtiar, S. Moradi, Estimation of asphaltene precipitation from titration data: a hybrid support vector regression with harmony search. Neural Comput. Appl. **26**(4), 789–798 (2015)

22. D. Manjarres, I. Landa-Torres, S. Gil-Lopez, J. Del Ser, M.N. Bilbao, S. Salcedo-Sanz, Z.W. Geem, A survey on applications of the harmony search algorithm. Eng. Appl. Artif. Intell. **26**(8), 1818–1831 (2013)
23. A. Assad, K. Deep, Applications of harmony search algorithm in data mining: a survey, in *Proceedings of Fifth International Conference on Soft Computing for Problem Solving*. Springer, pp. 863–874 (2016)
24. C. AoA, Official methods of analysis. Association of Official Analytical Chemists, Arlington, VA, USA **1**, 684 (1990)
25. Y.-H. Kim, Y. Yoon, Z.W. Geem, A comparison study of harmony search and genetic algorithm for the max-cut problem. Swarm Evolution. Comput. (2018)
26. M.N. Ambia, H.M. Hasanien, A. Al-Durra, S. Muyeen, Harmony search algorithm-based controller parameters optimization for a distributed-generation system. IEEE Trans. Power Delivery **30**(1), 246–255 (2015)
27. R. Eberhart, J. Kennedy, A new optimizer using particle swarm theory, in *Proceedings of the Sixth International Symposium on Micro Machine and Human Science, 1995. MHS'95*. IEEE, pp. 39–43 (1995)
28. S. Fong, R. Wong, A. Vasilakos, Accelerated pso swarm search feature selection for data stream mining big data. IEEE Trans. Serv. Comput. **1**, 1–1 (2016)
29. A.A. Aburomman, M.B.I. Reaz, A novel svm-knn-pso ensemble method for intrusion detection system. Appl. Soft Comput. **38**, 360–372 (2016)
30. M. Salucci, L. Poli, N. Anselmi, A. Massa, Multifrequency particle swarm optimization for enhanced multiresolution gpr microwave imaging. IEEE Trans. Geosci. Remote Sens. **55**(3), 1305–1317 (2017)
31. O.U. Rehman, S.U. Rehman, S. Tu, S. Khan, M. Waqas, S. Yang, A quantum particle swarm optimization method with fitness selection methodology for electromagnetic inverse problems. IEEE Access **6**, 63 155–63 163 (2018)
32. W. Yu, Z. Zeng, B. Peng, S. Yan, Y. Huang, H. Jiang, X. Li, T. Fan, Multi-objective optimum design of high-speed backplane connector using particle swarm optimization. IEEE Access, **6**, 35 182–35 193 (2018)
33. Y.-J. Gong, J.-J. Li, Y. Zhou, Y. Li, H.S.-H. Chung, Y.-H. Shi, J. Zhang, Genetic learning particle swarm optimization. IEEE Trans. Cybernet. **46**(10), 2277–2290 (2016)

Meme-Based Computational Optimization Framework

Felis Dwiyasa, Meng-Hiot Lim, Ren-Xiang Foo, and Shi-Wei Jason Teo

Abstract From a computing perspective, a meme denotes information that represents knowledge, patterns, rules, or strategies used to solve complex problems. When applied on a problem, memes help a solver to arrive at good quality solutions more efficiently, guiding the search process according to certain procedures or rules, instead of randomly searching through the solution space. Depending on the complexity of the problems, evaluating the suitability of memes and selecting a set of effective memes for different problems, however, are not straightforward tasks. A meme that works well for some problems may not be effective for other problems. Besides, different memes might have different degrees of importance in solving a problem. The level of importance of each meme might also change at different stages of the search. In this paper, we discuss how multiple memes can be generated and applied to solve computational optimization problems. A case study in combinatorial optimization is also presented and discussed.

1 Introduction

In the field of artificial intelligence, various learning paradigms have been introduced over decades. Regardless of the differences between various learning paradigms, learning always involves some information, knowledge, patterns, rules, or strategies that change the behavior of the learner.

The science and the art of selecting and incorporating memes within an evolutionary computing framework is the fundamental pillar of a memetic computational problem-solving paradigm (MCPP). Instead of searching for a solution from zero information, a MCPP capitalizes and exploits prior knowledge to guide the direction of the search. Each agent has access to a meme pool comprising various memes. Memes can also be passed on to another agent through propagation, selection, and variation mechanism [2, 12, 13]. MCPP has been used to solve various problems,

F. Dwiyasa (✉) · M.-H. Lim · R.-X. Foo · S.-W. J. Teo
Nanyang Technological University, Singapore, Singapore
e-mail: fdwiyasa@ntu.edu.sg

© Springer Nature Singapore Pte Ltd. 2020

A. K. Nagar et al. (eds.), *Soft Computing for Problem Solving 2019*,
Advances in Intelligent Systems and Computing 1139,
https://doi.org/10.1007/978-981-15-3287-0_12

such as multi-objective scheduling [18], vehicle routing problem [10], and inverse kinematics [17].

Despite the realization of memes as being the fundamental driver in complex problem-solving, memetic computing has often been simplistically alluded to as a form of hybridization. In its most basic and fundamental form, such hybridization is simply a combination of evolutionary algorithm and heuristic local search algorithm, otherwise referred to memetic algorithm or simply MA. Although hybrid evolutionary algorithm is a form of MA, its paradigm only fits the definition of the first generation of MA. The second generation of MA involves a pool of memes that go through selection and transmission mechanism. Through these mechanisms, memes that perform well will thrive, whereas memes that do not perform well will diminish over time. In the third-generation MA, the memes themselves can evolve into other types of memes.

This paper is aimed at providing a computational optimization framework that is based on memes as being part of the evolvable and transferrable entities during the search. The memes are not limited to local search heuristics, but also other problem-specific information, knowledge, patterns, rules, or strategies. In addition, we propose how multiple memes can be incorporated within a MCPP.

The rest of this paper is organized as follows. Section 2 presents different ways of incorporating multiple memes into a search algorithm. In Sect. 3, we propose a systematic classification of memes that allows meta-learning framework to generate new memes from basic memes. Section 4 presents a case study of combinatorial optimization, which describes some commonly used basic memes and illustrates how new memes can be constructed. Finally, we present the conclusions and some future works in Sect. 5.

2 Multiple Memes in Evolutionary Algorithms

Meme is a unit of information, knowledge, patterns, rules, or strategies incorporated within an agent. The first generation of MA is a simple hybrid evolutionary algorithm and local search meme, whereas the next generations of MA involve evolution process on the memes [11].

Meme-based optimization algorithms may rely on either a single meme or multiple memes. Figure 1a shows a simple hybrid algorithm where a single local search meme is applied at each generation. MA with a single meme is relatively easy to interpret and requires less computational resources. However, a single meme may not be enough to provide exploratory or exploitative orientation in the search process, resulting in ineffective convergence toward reasonable solutions. With multiple memes, the diverse range of the memes offers different paradigms on how to guide the search process, producing solutions with better quality as compared to a single meme. The term "multiple memes" in this paper is not to be confused with multi-meme MA proposed by Krasnogor et al. [8]. Here, we use the term to describe a rich set of

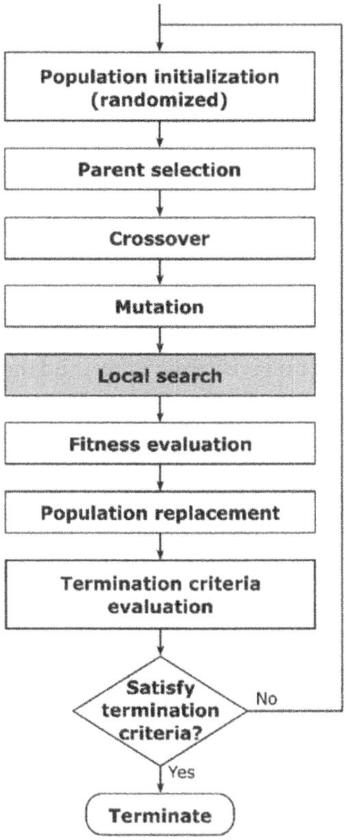

(a) With a single local search meme

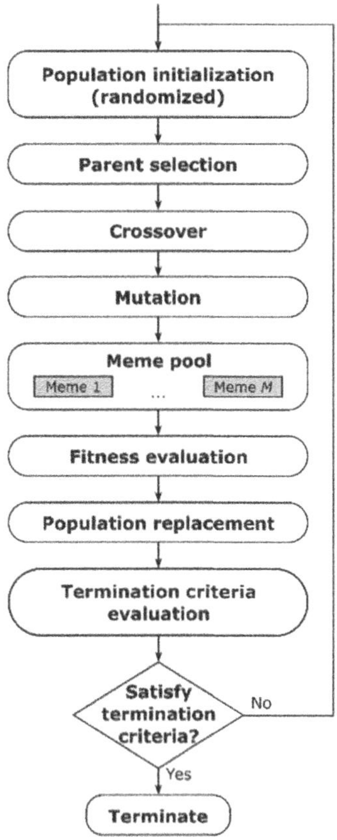

(b) Incorporation of multiple memes

Fig. 1 Computational framework—single meme versus multiple memes

memes that can be used to guide MA as shown in the meme-based computational framework illustrated in Fig. 1b.

Search algorithms with multiple memes are more complicated than the ones with single meme. There are also some design considerations on how the memes should be managed, as summarized in Fig. 2.

We may apply the memes one-by-one at each generation or altogether simultaneously. In between those two extremes, we may partially select some memes to be applied at each generation. Applying one meme at a time is straightforward and faster to compute, but the solution quality improvement is expected to occur at a slower rate. For M memes, if there is only one meme applied at each generation and each meme is given the same chance to be applied, the utilization rate of each meme would be $1/M$. Applying several memes simultaneously, on the other hand, increases the utilization rate of the memes. However, when applied simultaneously, the memes

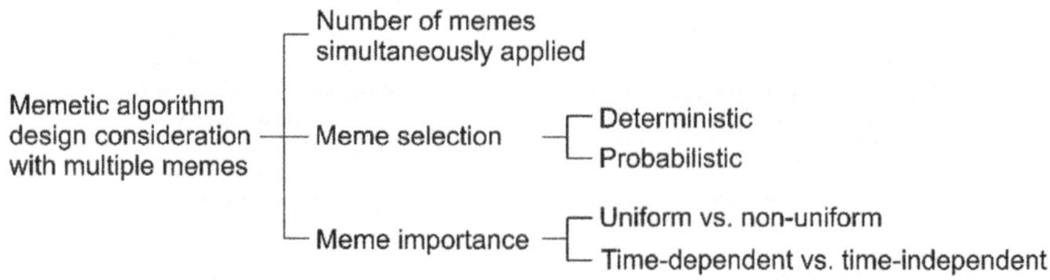

Fig. 2 Design consideration of memetic algorithm with multiple memes

combined could have a positive or negative interaction with each other, and thus, the higher utilization rate does not guarantee faster improvement of the solution quality.

Selecting a meme or a set of memes to be applied at each generation can be done using deterministic or probabilistic method. Some possible deterministic meme selection methods are random selection, round-robin procedure, weighted round-robin, and priority-based selection. As for probabilistic method, the memes are stochastically selected, where the probability of selecting each meme can be assigned a priori or obtained through meta-meme.

One of the most complicated issues involving multiple memes is deciding what memes are important and beneficial to be included. Although one may logically think of certain ways to solve a problem, it may turn out that it is not the best option available out there. As for ineffective memes, although the presence of them may not have a negative impact on the solution quality since inferior solutions are typically eliminated through selection biases, those memes would incur computational resources. Ideally, memes that are more important or effective should be applied more frequently, whereas the less important memes are applied less often.

The degree of importance might also change at different stages of the search. Some memes work more effectively at the beginning of the search or toward the end of the search, while some others are not affected by the stage of the search. Based on the characteristic of the degree of importance, we define the former type of memes as time-dependent and the latter ones as time-independent. The degrees of importance that are time-dependent would have a state variable that is determined by a certain time-dependent parameter.

3 Memes Generation Through Meta-Meme Framework

The concept of meta-meme, which is "beliefs about beliefs" or "ideas about ideas," can be traced back to the work presented by Hales [4], which conceptualizes that an agent can have a certain perception about the perception of another agent and act according to that perception. This concept is intriguing since it models higher-order intelligence that resembles how living things perceive knowledge and creates new knowledge. However, not many related works have been done to bring this concept closer into reality. Meuth et al. [11] propose higher-order learning framework that contains a set of optimizer, memory, selector, and generalization mechanism to generate new memes. Song et al. [16] use neural networks to map problem features into control parameters to choose memes from a meme pool.

To date, MCPP implemented nowadays usually involves a static meme pool, where the set of memes in the meme pool are formalized before running the algorithm. Although the memes can be automatically selected [8] and the parameters of the memes can be self-adjusted [14], the memes in the meme pool themselves do not change.

Hypothetically, a true memetic approach should be endowed with the capacity to create new memes or having "memes that create memes." To some extent, this is similar to the concept of meme complexes (memeplex) introduced by Chen and Ong [3], where the associations or synergies between memes continuously update and affect the probability of selecting a meme after another meme. The rationale behind the work is that consecutive activations of multiple memes might produce a unique effect that cannot be obtained when executing the memes individually. However, memeplex proposed by Chen and Ong [3] solely puts emphasis on the consecutive activations of the existing memes rather than the generation of new memes.

We believe that the key to generate new memes lies in the ability to construct a complex meme from fragments of memes according to pre-established rules. This principle applies in language, for instance, where syllables construct words, words construct clauses, clauses construct sentences, and sentences construct paragraphs. The construction of the language requires categorization of words, also known as the parts of speech, where words have different roles in a sentence. Guided by grammar rules, words can be constructed into coherent sentences to communicate ideas.

Here, we provide an example of classification of memes in computational optimization shown in Fig. 3. This classification enables memetic algorithm to perform a mix-and-match procedure and create new variants of memes.

Basic meme refers to the simplest unit of meme, whereas complex meme refers to a meme variant that involves two or more basic memes. This is analogous to the categorization of sentence structures, where a simple sentence contains one clause and a complex sentence contains two or more clauses. Basic meme can be categorized further into heuristic meme, context meme, and conditional meme.

Heuristic meme contains a procedure that defines how to build or improve a solution. There are two types of heuristic meme: (1) constructive heuristic and (2) local search heuristic. Constructive heuristics start with an empty solution and builds

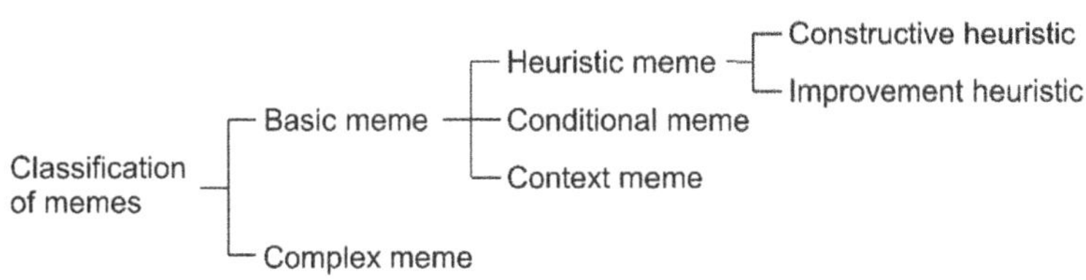

Fig. 3 Classification of memes

the solution step-by-step until a complete solution is constructed. On the other hand, local search heuristics or improvement heuristics require a complete solution as a start and improves the solution by applying local changes according to a certain procedure.

Conditional meme is a meme that needs to be fulfilled before another meme can be applied. This type of meme cannot be applied on its own and must be attached to a heuristic or context meme. Suppose **X** is a conditional meme and **Y** is a heuristic or context meme, a complex meme that contains a conditional meme can be formulated as "If **X**, apply **Y**." With conditional memes, improvements can be specifically targeted on problematic areas rather than applied across the whole search space.

Context meme is a higher-level information that describes the expected characteristics of the problem or the solution. Some examples of context meme are acceptance strategies, search intensity, cluster information, and recurring patterns. Compared to heuristics, this type of meme holds higher-level information that assists the search process to produce relevant solutions.

We revisited the architecture of higher-order meta-learning proposed by Meuth et al. [11] where the higher-order meta-learning consists of several lower-order meta-learning blocks, recursively down to the first-order learning. With this architecture, we applied the principle of memes classification to allow new memes for computational optimization to be generated and evolved.

Suppose we have the first-order meta-learning components containing the basic memes of each type in the memory elements. The second-order meta-learning would construct complex memes from the basic memes that are optimized and generalized for the problem. In the third-order meta-learning, the complex memes produced by the second-order meta-learning can be further used to construct the more complex memes. The orders of learning can continue to increase up to Nth order with the increase of the problem complexity and the required level of abstraction of memes.

4 Combinatorial Optimization Case Study

Combinatorial optimization aims to find the best solution of combinatorial mapping from a finite set of objects. Some examples of combinatorial optimization problems are traveling salesman problem (TSP), vehicle routing problem (VRP), bin packing problem, and flow shop scheduling problem. Combinatorial optimization problems are encoded as integer permutation representation, where each gene in a chromosome in evolutionary algorithms contains a unique integer number.

Given a set of cities, the objective of TSP optimization is to find the sequence of cities to be travelled by a salesman such that the overall cost of visiting all the cities is minimized. The classical TSP specifies that each city must be visited only once, and the salesman starts and ends the trip at the same city. Figure 4 illustrates the typical GA chromosome representation of TSP, where each unique integer number represents a city. In this example, the salesman would start from city number 4 then go to city number 6, 2, 9, 7, 5, 1, 10, 8, 3, and head back to city number 4.

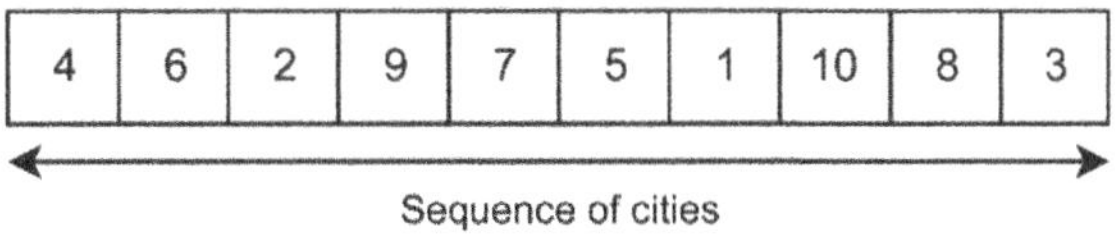

Fig. 4 Chromosome representation of TSP

Both constructive heuristics and improvement heuristics can be used to solve TSP. We can use constructive heuristics to generate the initial chromosomes and apply improvement heuristics to improve the solution quality at each generation. The use of constructive heuristics for population initialization has been found to improve the solution quality tremendously as compared to random initialization [15].

As for improvement heuristics, one of the most commonly used heuristics for TSP is Lin–Kernighan (LK) or k-opt method proposed by Lin and Kernighan [9], which is a generalization of 2-opt and 3-opt. The k-opt operator swaps some edges and selects the combination of edges that result in the shortest tour. According to performance tests conducted by Helsgaun [5], the solution quality improves with higher k, but choosing k higher than 4 is not recommended due to its insignificant improvement in the solution quality at the cost of a higher computational complexity.

Table 1 illustrates the chromosome representation and tour visualization for 2-opt heuristics, where 2 edges are swapped to construct a new tour. Since there are only 2 edges to be swapped in 2-opt, there is only 1 possible alternative solution to be compared. If the alternative solution has a shorter tour length than the original solution, the alternative solution will be used in the next generations.

For 3-opt heuristic, there are seven distinct combinations of tours after edge swapping [6]. As a result, the computational complexity of 3-opt is higher than 2-opt, but in principle, it can improve the solution quality faster. The chromosome representation and tour visualization of 3-opt are illustrated in Table 2.

The conventional way to apply k-opt heuristic is to systematically apply it across the whole search space or partially. Applying k-opt heuristic across the whole search space would undoubtedly produce a good solution, but the complexity increases exponentially with the problem size. Besides, after applying k-opt repeatedly, there will be a state when it is not able to improve the solution quality any further. If the

Table 1 Illustration of 2-opt heuristic for TSP

2-opt	Chromosome representation	Tour
Before	4 6 2 9 7 5 1 10 8 3	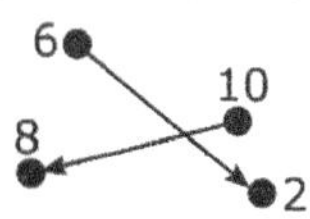
After	4 6 10 1 5 7 9 2 8 3	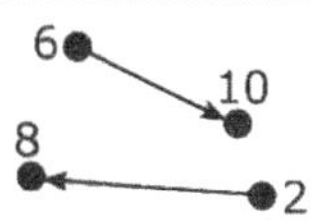

Table 2 Illustration of 3-opt heuristic for TSP

3-opt	Chromosome representation	Tour
Before	4 6 2 9 7 5 1 10 8 3	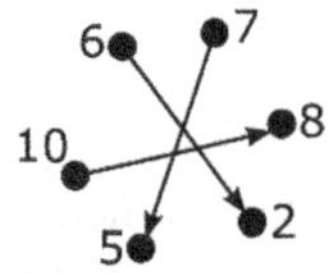
After	4 6 10 1 5 2 9 7 8 3	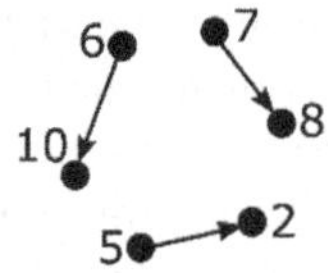

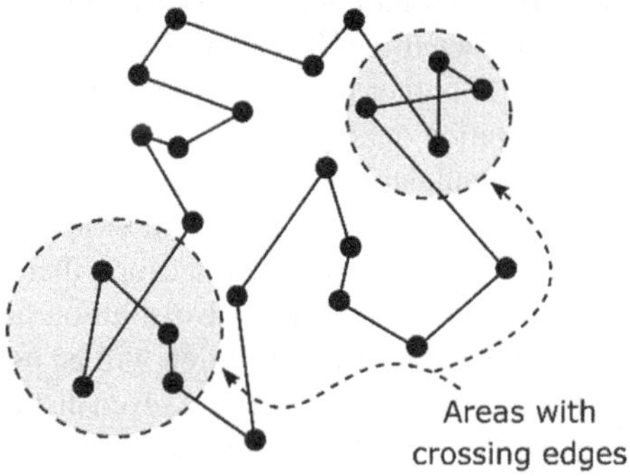

Fig. 5 TSP with the presence of crossing edges

heuristic is supplemented with a meme that could sense problematic areas that can be improved, the search process would be much more efficient.

To illustrate how the complex memes can be more efficient, consider a TSP tour illustrated in Fig. 5 and three options of memes illustrated in Fig. 6, where Meme A contains a basic meme and Meme B and Meme C contain complex memes.

Multiple traveling salesmen problem (MTSP) is a more generalized form of TSP, where the number of salesmen can be more than one. The typical objective function of MTSP is to balance the workload between salesmen or to minimize the total distance travelled [1]. MTSP is more complex than TSP since we need to determine both the assignment and the optimal sequence of the cities to be travelled by each salesman.

There are various chromosomal representations for MTSP, such as one-chromosome, two-chromosome, two-part chromosome techniques [1], and multi-chromosome technique [7]. In this paper, we use a one-chromosome representation as shown in Fig. 7.

Although TSP chromosome and MTSP one-chromosome representations look similar, applying k-opt heuristic is not effective to improve routes that belong to different salesmen in MTSP. A more effective way is to apply different heuristics

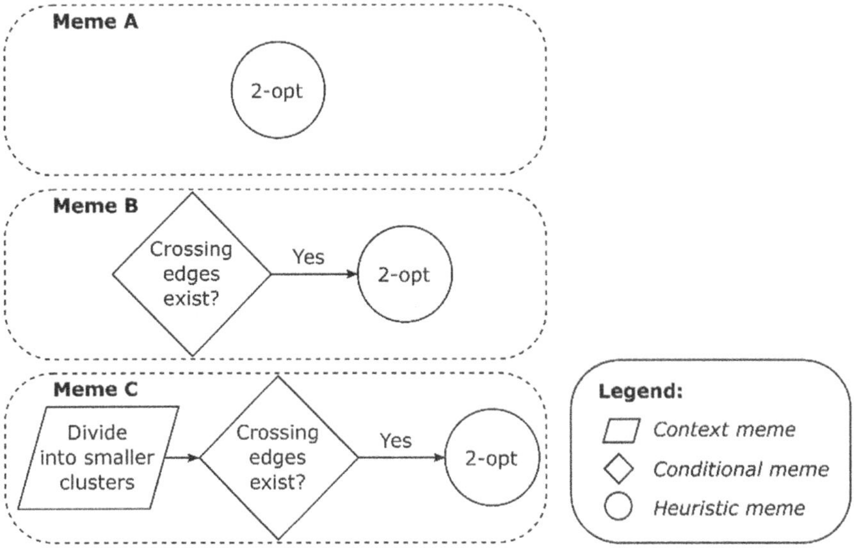

Fig. 6 Alternative options of TSP memes

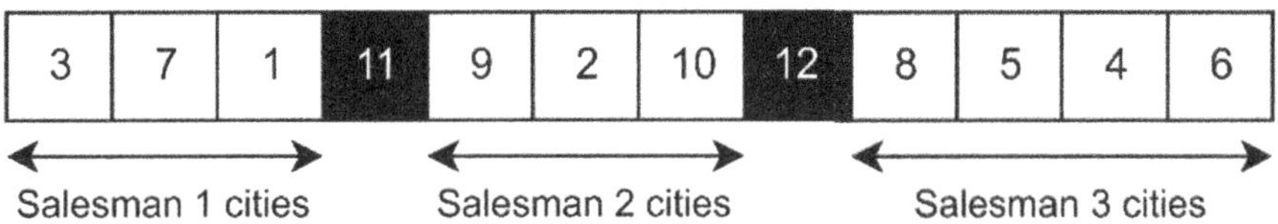

Fig. 7 Chromosome representation of MTSP

for in-route and cross-route [7], where in-route operator and cross-route operator are applied to improve the sequence of cities within one salesman and the routes of different salesmen, respectively. The cross-route operator used in the work is k-insert heuristic as illustrated in Table 3, where a part of the route of one salesman is cut and inserted into the route of another salesman.

Rather than exhaustively applying k-insert heuristic, we can construct a complex meme, named as Meme E in Fig. 8, containing k-insert heuristic and a context meme that finds the sections of the two routes that are nearby. Meme D illustrates the construction of a higher-order meme, where in-route condition is checked before applying Meme C for TSP in Fig. 6.

We may also generate some new variants of memes by simply changing one of the basic memes or the structure of the meme. Meme D, for instance, can be recreated into a new variant by changing the 2-opt into 3-opt, 4-opt, or other heuristic memes. The conditional meme that divides the problem into smaller clusters can also be differentiated by applying different clustering techniques.

Table 3 Illustration of k-insert heuristic for MTSP

k-insert	Chromosome representation	Tour
Before	3 7 1 11 9 2 10 12 8 5 4 6	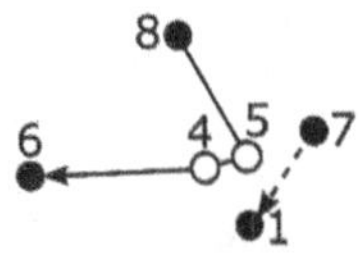
After	3 7 5 4 1 11 9 2 10 12 8 6	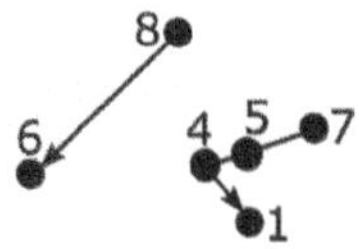

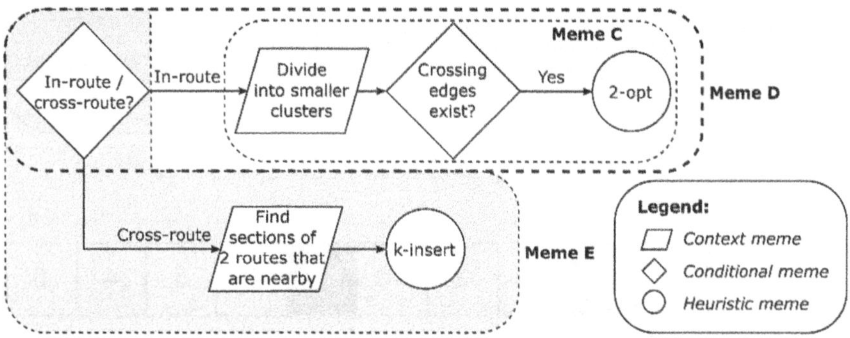

Fig. 8 Example of MTSP memes

5 Conclusions

A flexible framework for managing and incorporating memes into a search algorithm can uncover a whole new dimension of how computational optimization can be scaled up accordingly from the algorithm perspective. Rather than having the solutions generated through random evolutionary process or through a fixed set of memes, higher-level meta-learning architecture can be used to generate memes that might be more suitable to solve the problem at hand.

We illustrated the idea of memes generation in combinational optimization by showing how complex memes can be generated from basic memes. In principle, such memes are characteristically different and thus can be categorized. The compound memes of different categories can be further built into high-level complex memes. This demonstrates different levels of abstraction of memes and the possibility of creating new knowledge from some prior knowledge in memetic computation.

References

1. A.E. Carter, C.T. Ragsdale, A new approach to solving the multiple traveling salesperson problem using genetic algorithms. Eur. J. Oper. Res. **175**(1), 246–257 (2006)
2. X. Chen, Y.-S. Ong, M.-H. Lim, K.C. Tan, A multi-facet survey on memetic computation. IEEE Trans. Evol. Comput. **15**(5), 591–607 (2011)
3. X. Chen, Y.S. Ong, A conceptual modeling of meme complexes in stochastic search. IEEE Trans. Syst. Man Cybern. Part C (Applications and Reviews), **42**(5), 612–625 (2012)
4. D. Hales, Modelling meta-memes, in *Simulating Social Phenomena* (Springer, Berlin, 1997) (pp. 365–384)
5. K. Helsgaun, *An Effective Implementation of K-Opt Moves for the Lin-Kernighan TSP Heuristic*. Roskilde University. Department of Computer Science (2006)
6. H. Ismkhan, K. Zamanifar, Using ants as a genetic crossover operator in GLS to solve STSP, in *Paper Presented at the 2010 International Conference of Soft Computing and Pattern Recognition* (2010)
7. A. Király, J. Abonyi, A novel approach to solve multiple traveling salesmen problem by genetic algorithm, in *Computational Intelligence in Engineering* (Springer, Berlin, 2010) (pp. 141–151)
8. N. Krasnogor, B. Blackburne, E.K. Burke, J.D. Hirst, Multimeme algorithms for protein structure prediction, in *Paper Presented at the International Conference on Parallel Problem Solving from Nature* (2002)
9. S. Lin, B.W. Kernighan, An effective heuristic algorithm for the traveling-salesman problem. Oper. Res. **21**(2), 498–516 (1973)
10. Z. Meng, J.-S. Pan, Monkey king evolution: a new memetic evolutionary algorithm and its application in vehicle fuel consumption optimization. Knowl.-Based Syst. **97**, 144–157 (2016)
11. R. Meuth, M.-H. Lim, Y.-S. Ong, D.C. Wunsch, A proposition on memes and meta-memes in computing for higher-order learning. Memetic Comput, **1**(2), 85–100 (2009)
12. Y.-S. Ong, M.-H. Lim, N. Zhu, K.-W Wong, Classification of adaptive memetic algorithms: a comparative study. IEEE Trans. Syst. Man Cybern. Part B (Cybernetics) **36**(1), 141–152 (2006)
13. Y.-S. Ong, M.H. Lim, X. Chen, Memetic computation—past, present and future. IEEE Comput. Intell. Mag. **5**(2), 24–31 (2010)
14. E. Özcan, J.H. Drake, C. Altıntaş, S. Asta, A self-adaptive multimeme memetic algorithm co-evolving utility scores to control genetic operators and their parameter settings. Appl. Soft Comput. **49**, 81–93 (2016)
15. P.V. Paul, N. Moganarangan, S.S. Kumar, R. Raju, T. Vengattaraman, P. Dhavachelvan, Performance analyses over population seeding techniques of the permutation-coded genetic algorithm: an empirical study based on traveling salesman problems. Appl. Soft Comput. **32**, 383–402 (2015)
16. L.Q. Song, M.-H. Lim, Y.-S. Ong, Neural meta-memes framework for managing search algorithms in combinatorial optimization, in *Paper Presented at the IEEE Workshop on Memetic Computing (MC)* (2011)
17. S. Starke, N. Hendrich, J. Zhang, Memetic evolution for generic full-body inverse kinematics in robotics and animation. IEEE Trans. Evolut. Comput. (2018)
18. Y. Yuan, H. Xu, Multiobjective flexible job shop scheduling using memetic algorithms. IEEE Trans. Autom. Sci. Eng. **12**(1), 336–353 (2015)

References

[illegible]

Heterogeneous Multi-robot Mission Planning for Coordinated Tasks Execution

Felis Dwiyasa, Meng-Hiot Lim, Pyo Kang, Ren-Xiang Foo, and Shi-Wei Jason Teo

Abstract The use of robotics and autonomous systems to perform various missions has become increasingly popular and widely accepted. The scope of missions that can be performed by such systems includes complex tasks that require coordination and cooperation between multiple robots, not just simple repetitive tasks. Mission planner plays a vital role in making sure that the tasks performed by the robots are conducted in an efficient manner and meet all the mission requirements. In designing a mission planner, it is crucial to consider not only the functional aspect of the system but also its modularity, scalability, and interoperability. In this paper, we proposed an architecture of a heterogeneous multi-robot mission planner system that is based on robotic operating system (ROS). The system is designed to support heterogeneous multi-robot, where there are several types of robots having unique sets of capabilities and properties, which contributes greatly to the complexity of the mission planner.

1 Introduction

Various sectors such as agriculture, manufacturing, and logistics have been transformed and advanced by the emergence of robotics and autonomous systems in the past few decades. Unlike human who tends to make errors when carrying out tedious tasks over a long period of time, robots can perform such tasks efficiently. Tasks that may pose dangers to human, such as those dealing with hazardous substances or executed in harsh environments, can also be performed by robots.

Robots typically have some access to computational resources, which can be embedded in the robots themselves or located in a centralized server or in the cloud. These resources can be utilized to decide how the mission is going to be accomplished, which can be subject to some optimization criteria such as minimizing the time required to complete all tasks, minimizing the fuel usage, maximizing the usage of available resources, attempting to balance the workload, and many more.

F. Dwiyasa (✉) · M.-H. Lim · P. Kang · R.-X. Foo · S.-W. J. Teo
Nanyang Technological University, Singapore, Singapore
e-mail: fdwiyasa@ntu.edu.sg

© Springer Nature Singapore Pte Ltd. 2020

A. K. Nagar et al. (eds.), *Soft Computing for Problem Solving 2019*,
Advances in Intelligent Systems and Computing 1139,
https://doi.org/10.1007/978-981-15-3287-0_13

With the steady research progress in artificial intelligence, robots are now able to perform various types of complex tasks, not just simple repetitive tasks as in the past. Based on the level of interaction required to accomplish the tasks, complex tasks can be classified into two types: (1) loosely coordinated tasks, which are tasks that can be decomposed into several sub-tasks executed independently by different robots, and (2) tightly coordinated tasks, which cannot be decomposed and require strong coordination between robots during execution [7].

Multi-robot system (MRS) is similar with multi-agent system (MAS), with the exception that agents in MRS are physical robots that can interact in physical space, whereas agents in MAS are distributed stationary computers [9].

Based on their capabilities and properties, robots in MRS can be either homogeneous or heterogeneous. In homogeneous MRS, all the robots are equipped with identical actuators and sensing capabilities, whereas in heterogeneous MRS, each of them would have a unique set of capabilities and properties, making each robot suitable to handle specific types of tasks. In heterogeneous MRS, each task needs to be allocated to a robot that can fulfill the requirements of the task. MRS system assigns the tasks and schedules them in such a way that the overall execution cost is optimized.

When there is a massive number of tasks or when the tasks are complex, the system needs to apply task decomposition and task allocation. There are three different approaches known: decompose-then-allocate, allocate-then-decompose, or simultaneous allocation and decomposition approach. In decompose-then-allocate approach, task decomposition takes place before task allocation, allowing the task allocation process to be focused in optimizing a smaller number of tasks or a specific type of tasks. However, the resulted plan would be suboptimal since the task decomposition would not perform well without considering how the tasks are going to be allocated to the multiple robots. In allocate-then-decompose approach, task allocation takes place before task decomposition. This approach also produces a suboptimal plan because of the uncertainty of the costs of the allocated tasks, which are not exactly known before applying task decomposition. In simultaneous task allocation and decomposition, the robots simultaneously decompose the tasks into sub-tasks and trade the tasks or sub-tasks with one another [10].

In this paper, we presented the architecture of heterogeneous MRS that performs task decomposition and task allocation to handle loosely coordinated tasks, with the ability to perform task re-allocation in response to changing situation. We applied several algorithms, such as pathfinding and path optimization which involve heuristics and population-based metaheuristics for task allocation and re-allocation.

2 Problem Definition

There are two types of missions we considered in this work. The first type of mission is a coverage problem, which is often found in scenarios requiring surveillance and target detection. As illustrated in Fig. 1a, the robots in a coverage problem are required

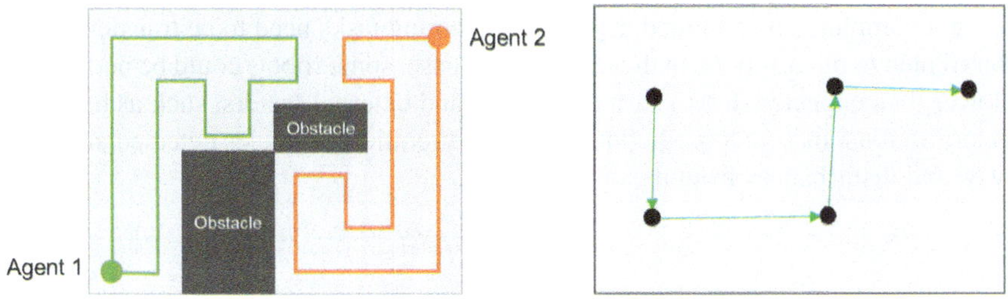

(a) Coverage (b) Route optimization

Fig. 1 Coverage and route optimization missions

to cover the whole area of a bounded region, where they perform certain sensing tasks or actions along the way. The second type of mission is route optimization, where the robots are required to follow a sequence of sensing tasks or actions at known locations as illustrated in Fig. 1b. Some examples of applications requiring a combination of these two missions are search-and-prosecute [1] and disaster relief [4].

In real-world scenarios, where the missions we need to solve are often complex in nature, heterogeneous robots with unique sets of capabilities are often more preferred than having one type of robot performing everything. Heterogeneous robots are usually more compact, cheaper, and easier to maintain.

In our work, we developed two types of robot prototypes, which are unmanned ground vehicles (UGVs) and unmanned aerial vehicles (UAVs). By having a wider angle from a higher altitude, UAVs can cover a wide area more effectively than UGVs, which are an essential feature for surveillance and target detection. On the other hand, UGVs can perform specific tasks on the ground that cannot be performed by UAVs. This illustrates how heterogenous robots with different capabilities can complement each other.

When performing a mission, robots must not come into the perimeters of objects and areas which can cause harm to the objects or the robots. Those objects and areas, which are considered obstacles, could be of different types for different types of robots. For example, water bodies must be avoided by the UGVs but not by the UAVs.

Locations of static obstacles can be loaded from a predefined map, if there is any, whereas the locations of moving obstacles such as human, animals, and other objects that are not available on the predefined map must be detected using real-time sensors. Obstacles that can only be detected in real-time are one of the causes why task re-allocation is needed.

We identified some possible scenarios that may cause deviation from the initial plan, at which the task re-allocation needs to take place. In a dynamic environment, new tasks or targets might emerge at any point of time, causing the initial plan to be outdated. When this occurs, the initial plan needs to be updated to include the new targets. Besides, during mission execution, there is a chance that a robot has malfunctioned or incapacitated due to unforeseen circumstances. If the robot has not completed its assigned tasks, the remaining tasks need to be transferred and distributed to other robots. In the less extreme case, some robots could be performing slower than expected due to various internal and external factors, such as hardware deteriorations, incorrect predictions of task difficulties, obstacles detected along the way, and disturbances from the environment.

3 System Architecture

We designed our heterogeneous MRS with a modular architecture, where the communication between each module is implemented using robotic operating system (ROS). As shown in Fig. 2, there are three key modules under ROS topic that are connected to the ROS master: mission description, mission planner, and hardware coordinator.

Mission description provides a graphical user interface (GUI) that allows a human user to define a mission and trigger the task allocation as well as the execution of the plan. When the mission is being executed, the user can monitor the execution progress and be alerted when there is any condition that might demand a task re-allocation.

Mission planner generates a plan based on the mission requirements that has been defined by the user via GUI. It performs two stages of computation: path planning and optimization. During the path planning stage, mission planner computes the distances between every pair of points while considering obstacles that might present between the points. The distance values are stored and ready to be used by the optimizer. Since there are different types of robots, we need a separate cost computation for each type of robot. Then, the optimizer finds the allocation of the tasks and the sequences that minimize the total cost. The cost can be a single-objective cost, where the total

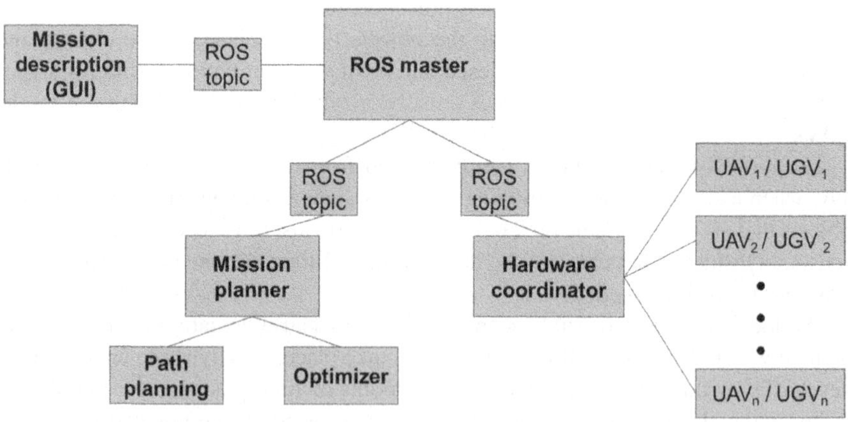

Fig. 2 System architecture

distance or execution time is the only factor, or a multi-objective cost concerning other factors such as fuel consumption or load balancing.

Hardware coordinator encapsulates the robot modules in such a way that every robot behaves similarly from the point of view of mission planner. Robots can be emulated using simulator or run on the real hardware. Due to safety considerations of the robot hardware, we allow users to choose manual control and autonomous mode anytime during mission execution. In case the autonomous mode goes out of control, the users can take over the robot control using manual mode. We use Ardupilot open-source autopilot firmware to enable autonomous control for both UAVs and UGVs, where the firmware parameters of the robots can be easily configured with a user-friendly interface.

4 Mission Planning

In real-life missions, producing suboptimal plans are often considered satisfactory and more practical than attempting to get a global optimal plan due to several reasons. Firstly, relaxing the optimality of the plan would save a significant amount of optimization time and computational resources, especially when the scale of the mission is massive. Secondly, the cost of executing the global optimal plan might only be slightly lower than the cost of executing a suboptimal plan. The insignificant cost reduction may not be worth the extra time needed for planning. Lastly, mission execution may not go as planned due to unforeseen circumstances, particularly because the conditions encountered at the time of execution might have changed from the conditions captured at the time of planning. If the plan is most likely going to be amended due to the dynamic nature of the mission, it is not necessary to have a global optimal plan to begin with.

Task allocation is a combinatorial problem that falls into the category of NP-hard problems. To solve NP-hard problems, exact algorithms are rarely used because of their limited scalability for large problems. Heuristics methods are fast and scalable, but they tend to be quickly trapped at local minima. Evolutionary algorithms or metaheuristics, on the other hand, are known for their ability to produce reasonable solutions within a much shorter time as compared to exact algorithms. Some examples of metaheuristics are genetic algorithm (GA) [5], particle swarm optimization (PSO) [8], ant colony optimization (ACO) [2], and simulated annealing (SA) [3].

Unlike exact algorithms and heuristics, search process in metaheuristics always involves some stochastic randomness. Typically, the process involves generating an initial population and repeatedly applying evolutionary mechanisms such as crossover and mutation operators. The randomness introduces diversity into the solution candidates, which is essentially the principal mechanism that prevents the final solution from being trapped at local minima. However, relying primarily on a random-search mechanism is time-consuming and might not produce the desired optimal solution, especially for large problems where the size of solution space is enormous.

As such, the incorporation of information either at the onset or progressively can enhance the capability of the search process.

Considering that we need to get a good solution quickly to fulfill the real-time requirement of the system, we use a memetic algorithm (MA) approach [6] which contains metaheuristics guided by memes. We use GA as the metaheuristics platform in this work. Since metaheuristics produce valid solutions from the beginning of the search, we can terminate the optimization process anytime according to the time budget that we have. Meanwhile, the search guidance given by the memes allows the solutions to improve faster at the beginning of the search, resulting in a good solution that we can use as early as possible.

As for the mid-mission task re-allocation, we implemented a local segment optimization that selectively reoptimizes path segments when there is changing situation during mission execution. Selective re-optimization is much faster than re-optimizing the whole plan. Furthermore, for most practical scenarios, drastic or significant changes from the initial plan are undesirable or impractical.

Computational time of task re-allocation during mission execution is more critical than during initial task allocation because of the real-time nature of the mission execution. If the computation of the new plan takes longer time than the mission execution, the new plan cannot be distributed to the respective robots in time, and the conditions would have changed from the time we initiated the task re-allocation. When this happens, the new plan we have generated would not be relevant anymore.

5 Conclusions

We have presented an architecture of a heterogeneous multi-robot mission planner system that is designed to support heterogeneous multi-robot, where there are several types of robots having unique sets of capabilities and properties. The main challenge we encountered is to optimize the task allocation while following the real-time requirements of the system. Therefore, we need to carefully select algorithms that can scale well with the problem size and provide good solutions within reasonable time.

References

1. T. Abukhalil et al., Coordinating a heterogeneous robot swarm using Robot Utility-based Task Assignment (RUTA), in *2016 IEEE 14th International Workshop on Advanced Motion Control (AMC)* (IEEE, 2016)
2. M. Dorigo, C. Blum, Ant colony optimization theory: a survey. Theor. Comput. Sci. **344**(2–3), 243–278 (2005)
3. K.A. Dowsland, J.M. Thompson, Simulated annealing. in *Handbook of Natural Computing* (2012), pp. 1623–1655

4. J. Gregory et al., Application of multi-robot systems to disaster-relief scenarios with limited communication. in *Field and Service Robotics*, (Springer, 2016)
5. J.H. Holland, Genetic algorithms. Sci. Am. **267**(1), 66–73 (1992)
6. Y.-S. Ong et al., Memetic computation—Past, present & future. IEEE Comput. Intell. Mag. **5**(2), 24–31 (2010)
7. Y. Rizk et al., Cooperative heterogeneous multi-robot systems: a survey. ACM Comput. Surv. (CSUR) **52**(2), 29 (2019)
8. Y. Shi, Particle swarm optimization: Developments, applications and resources, in *Proceedings of the 2001 Congress on Evolutionary Computation* (IEEE, 2001)
9. Z. Yan et al., A survey and analysis of multi-robot coordination. Int. J. Adv. Robot. Syst. **10**(12), 399 (2013)
10. R. Zlot, A. Stentz, Market-based multirobot coordination for complex tasks. Int. J. Robot. Res **25**(1), 73–101 (2006)

Development of Cost-Effective Endurance Test Rig with Integrated Algorithm for Safety

Emanuele Lindo Secco, Rashid Abdulrahman, Ian Felmeri, and Atulya K. Nagar

Abstract This paper presents a novel algorithm for controlling a new endurance test rig for the validation of a safety switch of the Cortech HealthCare Ltd. shower and commode chair. The paper outlines the design of the algorithm as well as the main concept and criteria together with the validation and testing methods. In order to validate the algorithm, a set of preliminary trials is also presented.

Keywords Safety · Biomedical devices · Smart devices · Human–device interaction

1 Introduction

Over the past century, industrial automation in a form of machines and robotics has been developed in different sizes and shapes for a variety of reasons such as to increase efficiency, collect accurate data, and to save money. Automated systems allow for improvements that benefit from consistent execution.

One of the aims of this scientific research is to develop an efficient rig which serves the purpose of testing safety switches in a daily life tool for elderly people, namely a toilet chair. This research seeks to overcome the difficulties experienced in the previous research, with this being achieved through the repetition and critical analysis of all researches on the field of study that the project is based within.

E. L. Secco (✉) · R. Abdulrahman · A. K. Nagar
Robotic Laboratory, School of Mathematics, Computer Science and Engineering, Liverpool Hope University, Liverpool L16 9JD, UK
e-mail: seccoe@hope.ac.uk

R. Abdulrahman
e-mail: 16007751@hope.ac.uk

A. K. Nagar
e-mail: nagara@hope.ac.uk

R. Abdulrahman · I. Felmeri
Cortech Healthcare Ltd., Isle of Man IM9 2AJ British Isles, UK
e-mail: ian@cortech-healthcare.com

© Springer Nature Singapore Pte Ltd. 2020

A. K. Nagar et al. (eds.), *Soft Computing for Problem Solving 2019*,
Advances in Intelligent Systems and Computing 1139,
https://doi.org/10.1007/978-981-15-3287-0_14

From a historical perspective, the Third Industrial Revolution, which began in 1969, allowed for the automation of the product line, due to the developments in the mechanical devices, and more largely in mechatronics and robotics, and the invention of programmable logic controls (PLCs). Today, the Fourth Revolution is building upon the previous industrial revolution, which is also referred to as 'Industrial 4.0' [1, 2]. According to Professor Schwab [3], Founder and Executive Chairman of the World Economic Forum, this Fourth Industrial Revolution is seen to differ to the previous industrialization, and it is concerned with the fusion of the physical, digital, and biological worlds to the extent which impacts all disciplines, economies, and industries. The Fourth Revolution has also allowed for the questioning of what it means to be human [1]. Industrial 4.0 is characterized as 'the development of AI, genome editing, 3D printing, and autonomous automobiles' [4].

Out of Industrial 4.0, the development and the emergence of mechatronics and robotics have become apparent in contemporary society. Over the five decades, research has been extended into the study of intelligent systems; typically, research serves a purpose to overcome specific technological issues in certain area, leading to advances in 'systems and solutions that will profoundly impact society' [5]. The justification of the robotic processes is based upon the premise that the process allows for 'safer operation, savings of materials, and significant savings in terms of man-hour and production line costs' [6–10].

In this context, we are proposing a novel integrated design which combines mechatronics elements with an intelligent algorithm in order to improve human daily life and specifically, the life of elderly people. Currently, Cortech HealthCare Ltd. does not have a testing facility that can be used to test the safety mechanism of the shower and commode chair [11]. The aim of this paper is to present the development of a cost-effective method that shall be used to test the safety mechanism on the shower and commode chair.

2 Materials and Methods

This section is used to discuss, present, and give a clear understanding of the methods and procedures used into developing the project. The section is divided into multiple parts that are going to be used throughout the project with subsections for each part: hardware, software, and implementation section. Each section shall have a brief explanation of how every element functions and how it shall be used within the project.

The hardware section is divided into controllers, sensors, actuators, and any other additional equipment used throughout to support the project.

The software section shall have a brief explanation of some of the software used to develop the project software such as SolidWorks 2018, Visual Studio 2018, Arduino IDE, and simple 3D shall have a description, and there purpose throughout the project

The last part is the implementation part where some of the above-described elements were implemented into the project. Some of the hardware parts such as mechanical parts shall have a description with diagrams on how they were constructed into the final product and other electronics parts shall pin connection table and layout diagram. However, for brevity we have not included them.

2.1 *Hardware*

2.1.1 Controller

For this project, an Arduino Mega 2560 shall be used as a controller. Arduino Mega is a microcontroller board that is constructed on ATMEGA 2560 chip which has a clock speed of 16 MHz. The board operates at 5 V and has 2 power sources; it embeds a USB and a DC power supply which can take an input voltage between 7 and 12 V recommended, but it can run between 6 and 20 V absolute limits. Arduino Mega 2560 has 54 digital I/O pins, 12 of which support pulse width modulation (PWM). They have internal pull-up resistance between 20 and 50 kH that are turn-on and -off by the software. The board also has 16 analog pins and 6 interactive pins (Fig. 1).

All 54 digital pins on the board operate at 5 V and can send or receive maximum of 40 mA. These pins can be utilized as input or output using commands/functions such as *pinMode()*, *digitalRead()*, or *digitalWrite()*, on top of that few pins consist of specialized functions such as serial communication (e.g., I2C, SPI).

Arduino Mega 2560 has 256 kB of flash memory and 8 kB of programmed memory as well as 4 kB of EEPROM which can be used to read and write with EEPROM library.

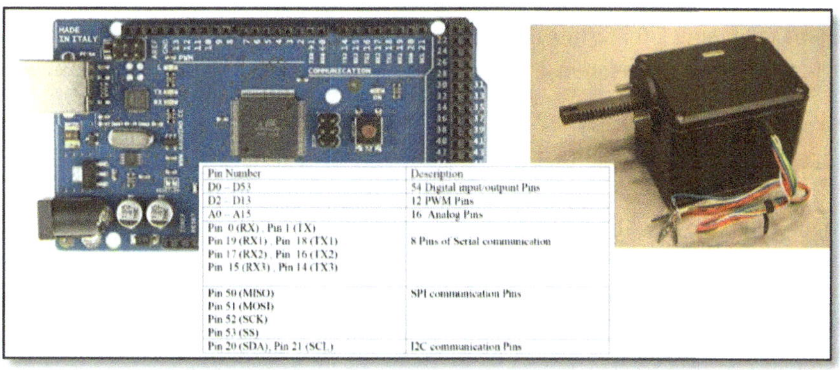

Fig. 1 Arduino Mega 2560 setup and pin configuration (left panel) and the A.R.A. linear actuator (right panel)

2.1.2 Actuator

A.R.A. linear actuators are going to be used in this project. This type of actuators was initially designed to be used as cruise control in agriculture vehicles, and these are very inexpensive actuators that have an electrical operated clutch that allows its mechanical arm to be disconnected from the system. These actuators associated with Arduino and L298N boards are ideal for home projects and automotive. This type of actuator also contains a potentiometer that allows the position of the linear arm to be sensed (Fig. 1).

The actuator is rated at 12 V, and the clutch consumes around 120 mA when engaged while the motor draws around 500 mA while engaged under reasonable heavy load. These actuators are 95 × 72 × 75 mm and have a stroke of 50 mm; the time travel of the arm from one end to the other is around 3 s. They consist of solid aluminum die-casting case with high-quality gear grain.

Three of these actuators shall be used with the system to measure three different points of sensitivity on the proposed test rig.

2.1.3 Servo Motor Controller

L298N is a dual full-bridge rectifier module that can be used to drive speed and direction of a servo or motor; it can also drive any other inductive load relays. L298N board can be driven manually using switches or much more efficiently using microcontrollers such as Arduino. This board has an input voltage value of between 5 and 35 V. This board was selected because it consists of two terminal blocks label as output A and output B as shown in Fig. 2 (part 1). One terminal shall be used to drive the actuator while other terminal shall be used to drive to engage or disengage the clutch in the actuator.

The L298N board has three logic inputs for each terminal labeled as enable A, input 1, and input 2 for terminal 1, and the second terminal has the same logic where there is an enable B, input 3, and input 4.

2.1.4 Load Cell Strain Sensor

Strain gauge sensor is the device that measures a load/strain of an object by measuring its resistance. This is due to its electrical conductance on conductor geometry. Strain gauges are used for many applications such as stress analysis, measuring forces such as pressure, acceleration, or displacements. When force is applied to a gauge, the area around the gauge narrows and its resistance increases; the variation in resistance of the gauge is reasonable linear function of its formation, and it transforms mechanical strain to electrical signal.

Load cells usually follow a Wheatstone bridge configuration layout where one-quarter bridge load cell is used or two half-bridge load cells are used. The electrical signal from these cells is usually within few mV and needs to be amplified before

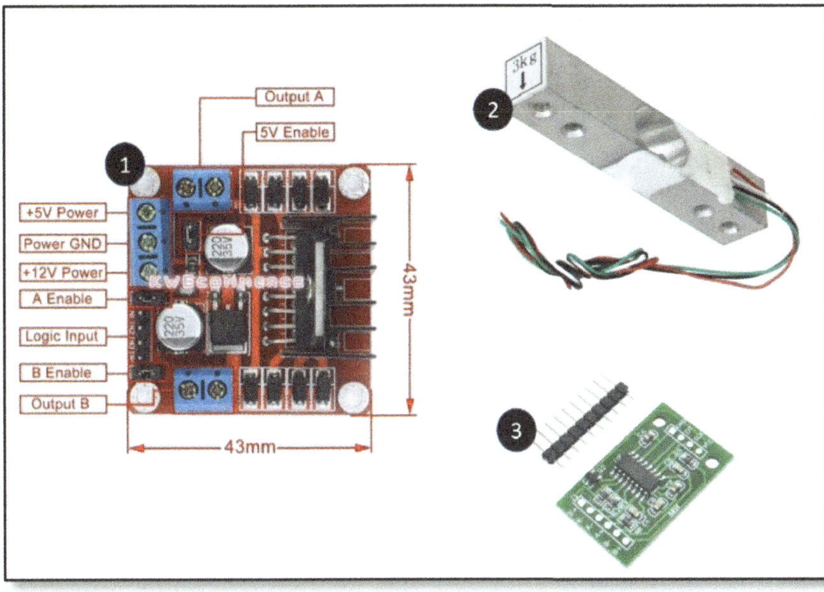

Fig. 2 L298N motor driver (part 1), the XYZ-133 Mini Scale sensor (part 2) and the HX711 module (part 3)

being used. The output of the transducer can be scaled to calculate the force applied to the transducer. Sometimes, a high-resolution ADC, typically 24-bit, can be used directly.

For this project, YZC-133 Mini Scale sensor is going to be used in cooperated with HX711 AD module (Fig. 2). These components were selected because they are cheap and easy to buy. YZC-133 Mini Scale sensor is also a full bridge load cell sensor and therefore can be used as a stand-alone unit.

2.1.5 HX711 Module

HX711 module is 24-high precision analog-to-digital converter. This module has two analog input channels and a programmable gain value of 128-integrated amplifier (Fig. 2). The input can be calibrated to provide a bridge voltage of pressure. Here are the main features:

- Two selectable differential input channels
- On-chip
  - active low noise PGA with selectable gain of 32, 64, and 128
  - power supply regulator for load cell and ADC analog power supply
  - oscillator requiring no external component with optional external crystal
  - power-on-reset

- Simple digital control and serial interface: pin-driven controls, no programming needed
- Selectable 10 or 80 SPS output data rate
- Simultaneous 50 and 60 Hz supply rejection.

This module shall be incooperated with the YZC-133 Mini Scale sensor, and it was selected because it has high precision, low-cost sampling.

A power supply is also integrated within the hardware system architecture together with some mechanical components such as a 20 × 20 series aluminum extrusion bars and a set of corner bracket gussets.

2.2 *Software*

2.2.1 3D Design Software

A 3D design software *SolidWorks*® (by Dassault Systèmes) was chosen. SolidWorks is a computer-aided design (CAD) modeling software that allows designers to design products and components in three dimensions. The procedure behind the software is generally to draft a 2D sketch and use lofting and extrusion to form a solid form. There are multiple advantages in using SolidWorks some being 3D modeling, test collaboration, and so on.

The SolidWorks 2018 version was chosen for this product because the operator can build and test majority of the features within the product to ensure that it meets the design and set criteria. This process could be used to save money and time on purchasing and fitting components that would not have been compatible with the design.

Figure 3 shows the Cortech HealthCare shower/commode chair; a similar design incooperated with an endurance test rig shall be used to test the safety mechanism of this product.

Simplify3D is a 3D printing slicer software that is capable of converting 3D drawings into set of printing instruction for a 3D printer. Simplify3D consists of multiple useful features for 3D printing such as realistic simulation that allows the operator to preprint simulation of a product and see the exact pattern the 3D printer takes. This process shows the exact animation that the extruder takes as it lays down the plastic. This simulation features other important information, such as sequence, and speed that can help identify any errors before wasting time into printing a failed part

Simplify3D was selected for this project because it supports our printer and its simulation feature can be used to identify errors before printing any parts.

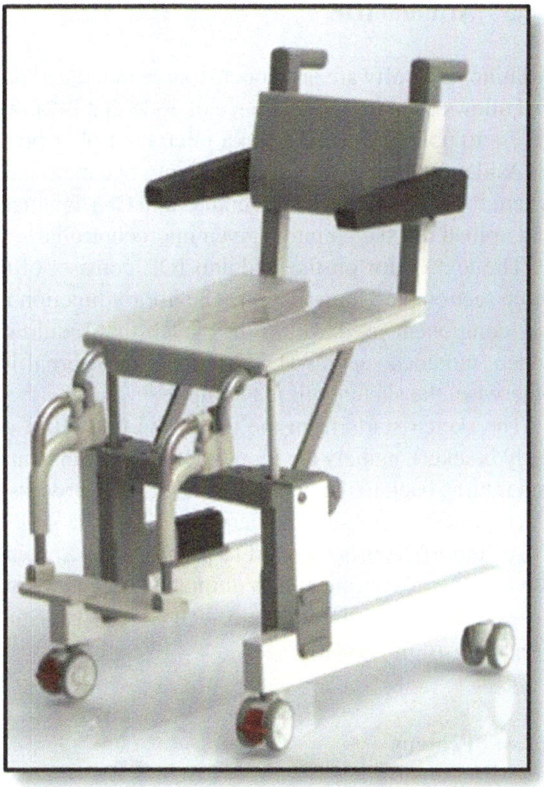

Fig. 3 Cortech HealthCare shower/commode chair

2.2.2 IDE Tool

Visual Studio® (by Microsoft Corporation) is an integrated development environment (IDE) tool developed by Microsoft. This tool consists of code editor supporting intelligence as well as code refactoring. It also consists of debugging tools that work as a machine-level debugger or source-level debugger. Other features included in the tool are the code profiler where it allows to form the designers for building GUI application, web designers, and so on. This IDE tool supports code editing and debugging for multiple programming languages such as C/C++, C#, .net and also supports other script languages such as ruby.NODE.js.

For this project, Visual Studio 2018 shall be used to develop a form and graphical user interface (GUI) that shall be used to send and receive signal from a controller via serial and also will be used to monitor the progress of the test. The GUI shall be written and edited in C#.

2.2.3 Arduino IDE

Arduino IDE software is an open-source integrated development environment (IDE) that allows users to write a piece of code that is known as sketch, and it gives them access to upload the code into a microcontroller board such as Arduino in form of HEX file. Arduino IDE is separated into two main parts, the editor and the compiler, where former is used to write both C and C++ languages and later is used to compile and upload the sketch into a given microcontroller.

The text editor on the Arduino IDE consists of three sections: library section, setup section, and loop section. The library function is at the top of the editor where any component used within the sketch that requires external files that need to be added, modules such as HX77 require an external library to be added before use; otherwise, the sketch will not compile.

The sketch starts from the top of the first bit of code (generally marked with a curly bracket), namely the loop function and runs until it reaches the closing bracket and returns back to the opening bracket. This process repeats itself until the board is reset, turned off, or stated otherwise.

Arduino IDE was selected to program the Arduino board because of its ease of use to write a program and to ultimately upload the written program to the Arduino board. Other software out there such as ATMEL Studio 7 can be used to write and program Arduino boards.

2.2.4 Proteus

Proteus is an electronic software that features schematic capture, printed circuit board (PCB) layout and simulation. Proteus simulation is capable of simulating any circuit that has a microcontroller such as the Arduino in real time; this is done by uploading a hex sketch that can be obtained from software such as Arduino IDE. This feature can be used to check for any errors on the sketch that has been written for the microcontroller.

Proteus shall be used in this project to test the C# GUI software that was written in Visual Studio 2018. This software minimizes the need of a physical Arduino, and it can therefore be used before building up the rig.

Proteus consists of vast list of build in components, and many libraries and sometimes other libraries can be downloaded and added into a component list: for example, component such as Arduino is not preloaded into the library, and therefore, it needs to be downloaded.

2.3 *Design, Fabrication, and Integration*

2.3.1 Design

After some preliminary work, it was decided that it will be the best fit to design a frame/rig that is portable and does not need to be fixed at a specific position and by doing so the test can be set up at any location within the office.

The first step into designing the rig was to decide the configuration of the test, and this was decided by inspecting the location of the safety switches of the chair and weight on the advantages and difficulties that might occur throughout the test. Accordingly, it was decided that it would be best fit to design a portable test rig that will fit around the bottom half of the tested chair.

Figure 4 (left panel) shows a mock-up design of the rig that was performed with SolidWorks 2: The rig configuration consists of three mains A.R.A linear actuators that shall move up and down directly on top of the tested chair according to an optimized intelligent algorithm.

Below the actuator, there is a case that shall hold a sensor that is used to measure the sensitivity of the safety mechanism before being triggered. The sensor case itself contains its own mechanism that was designed *to mimic a functionality of a finger* (Fig. 4, right panel).

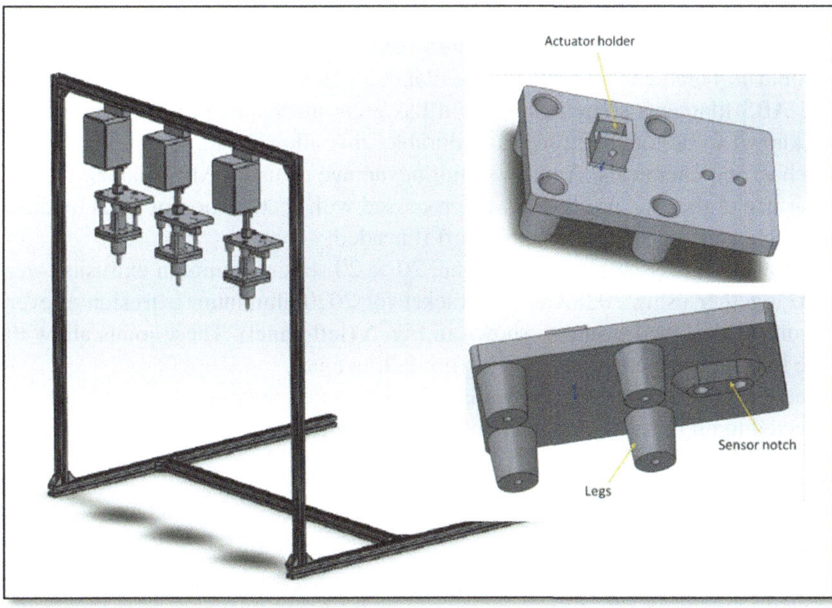

Fig. 4 Endurance test rig design (left panel) and the details of the strain gauge holder parts (right panel)

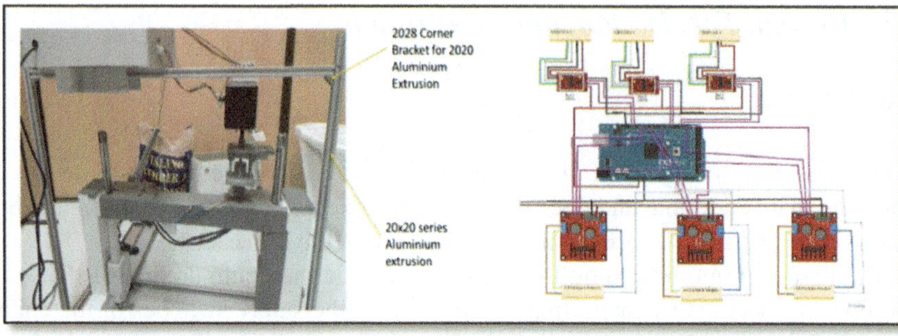

Fig. 5 Manufacturing of the endurance test rig frame (left panel) and the integrated electronics and circuit design (right panel)

The mechanism consists of two rams. One at the top and bottom connected together using a spring that is placed in the center between two rams, and the main aim to this type of mechanism is to give a sensor and the tested part of the seat a smooth force transaction other than just being hit or push with a solid component.

2.3.2 Fabrication and Components Integration

To reduce the cost and to keep the rig lightweight, the sensor case and some of the connector blocks were manufactured using raised pro2 plus 3D printer using acrylonitrile butadiene styrene (ABS) filament. There are multiple advantages in using ABS filaments some being that it has great mechanical properties, and it is well known to be tough, strong, and durable; this allows it to be tolerant of heat, scratches, and chemicals. Another major advantage of using ABS filaments is that the surface of the final product can be processed with acetone or glue parts together, and lastly, some material can be filed off if needed.

Following the assembly of the frame, 20 × 20 series aluminum extrusion were linked together using 2028 Corner Bracket for 2020 aluminum extrusion at every joint on the designed frame as shown in Fig. 5 (left panel). These joints allow the frame to be held together and stop it from falling apart.

The electronic components were also integrated within the rig frame and connected to all the hardware elements (Fig. 5, right panel).

3 Design of the Algorithm

In this section, all the design of the algorithm and written codes are reported. This contribution mainly refers to the aforementioned Arduino IDE software.

The overall main rational of the algorithm is based on performing the following sequence of action and check:

- Set up the test and move to start position (cycle start at 0)
- Start test by extracting actuator
- Observe switch
- Observe potentiometer position or limit
- When switch is triggered, record potentiometer position and applied force
- Retract actuator.

Accordingly, a flowchart was designed to give graphical representation of the algorithm. Flowchart allows us to easily create an algorithm and also allows us to analyze different processes within algorithms.

The schematics within the flowchart shown in Fig. 6 (left panel) displays the different parts of the algorithm, and throughout this process, we can highlight a certain element and relationship between each section of the algorithms.

In order to perform the task, the software needs to be uploaded with the proper libraries which link to the used different electronic components. In this context, the HX711.h library allows the Arduino board to communicate with the HX711 module which is connected to the load cell. The library also provides extra function that can be used in cooperation with the module, namely functions such as *.tare()*, *.read_average()*.

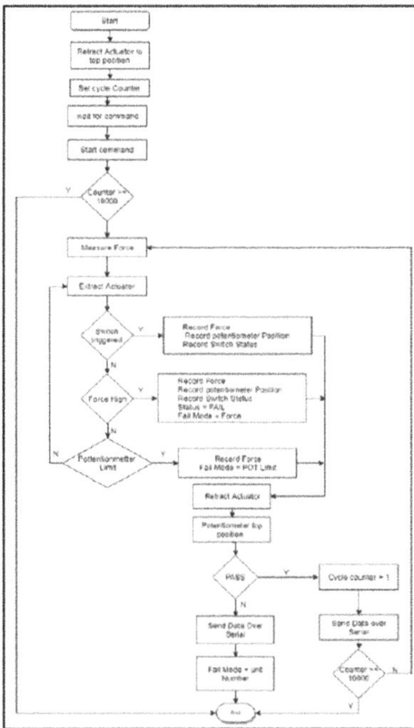

Functions		
Functions	Description	Called by
Retract ()	This function retract the actuator to its home position. this is done by first engaging the clutch and turning of signal to motor_e variable and sending signal to motor_r variable with power variable of 255	This function is called when case statement is set to 1 or 2
TestCycle ()	This function was used for checking any new added variable and there function without amending the core program in case 1 or 2	This function is called when colState is set to TestCycle
measure()	Measure function consists of load cell algorithm that is used to read a signal from the load cell	This function is called when extendedState is 1
force()	This function consists of fail measure for the load cell, once a set load has been exceeded the first part of this function will execute	This function is called when extendedState is 1

Fig. 6 Flowchart design (left panel) and the portfolios of the designed function (right panel)

Furthermore, a set of declared variables is introduced in order to define the output and input: This allows the Arduino module to determine if to send signal out or it need shall be receiving signal from the connected component.

Finally, a set of functions elaborates on all the processes used in the Arduino sketch. All the names of these functions used and how they are called are shown in Fig. 6 (right panel).

4 Results

A set of preliminary tests was performed in order to validate the integrated hardware and software systems.

4.1 Calibration

One of the main components in the design is the sensor, as this device is going to be used to measure the amount of load required to trigger the safety system on a shower commode chair and that was one of the aim reasons for designing and constructing the test rig. In this context, one of the main devices used in the rig is the YZC-133 Mini Scale sensor. Before placing the sensor on the rig, the sensors had to be tested to ensure that it is capable to perform at a reasonable level. For calibration and testing, the sensor was placed between two aluminum plates separated using two nuts and secured together using threaded bolts. Then a sequence of loading steps was performed: This was done by loading and unloading a known weight on top of the weighting plate and reading the output shown on the Arduino IDE. From these data, a weight of 0.74 kg was placed on the weighing plate. The produced output from this weight was 0.76 kg. The errors on the given output were adjusted to give a correct output by changing the sensor sensitivity using a *set_scale* function.

4.2 Hardware's Outcome

From the calibration results, it was noted that the sensor gave out a stable output with error tolerance of ±2. From these results, it was concluded that YZC-133 Mini Scale sensor can be used in the testing rig as it is capable of giving out stable results throughout the test. The output/tolerance of the sensor can be adjusted and improved by adjusting the units within the aforementioned *set_scale* function.

Following the experiments, a set of observations were performed. All the 3D printed parts in this rig were manufactured using ABS filament, and this type of material has its own drawbacks.

- *Temperature issues*: The ABS filaments require a good temperature management throughout the process, and any errors in the process will result in the final product to have cracks or split layers in them and will require the part to be reprinted. One of the ways to prevent this issue is to have all enclosures within the printer closed throughout the process.
- *Extruding problem*: Extrusion issues can happen at any time and cause major setbacks on printing time. No extrusion and under extrusion can be caused by variety of facts some being that the filament temperatures settings are incorrect, clogged nozzle, loose extruder gears. These issues most of the time cause the part to be reprinted and could delay the part even longer if the issue occurs near the end of printing or at the beginning of and someone failed to spot the issue.
- *Printing time*: Printing parts in high quality took a long time and sometimes the printer had to be set on print for more than 2 h and when something goes wrong with the printing, the entire process had to stop and start printing again from the start.

Among these problems, the spring mechanism itself shows some limits: From the carried out test run of the rig, it was concluded that there was not enough load being transferred to the YZC-133 Mini Scale sensor; this is due to the chosen spring not been stiff enough to transfer the load between the two rams in the unit (Fig. 7). This issue can be fixed by either replacing the spring to a stiffer string or change the design where the string is preloaded within the case.

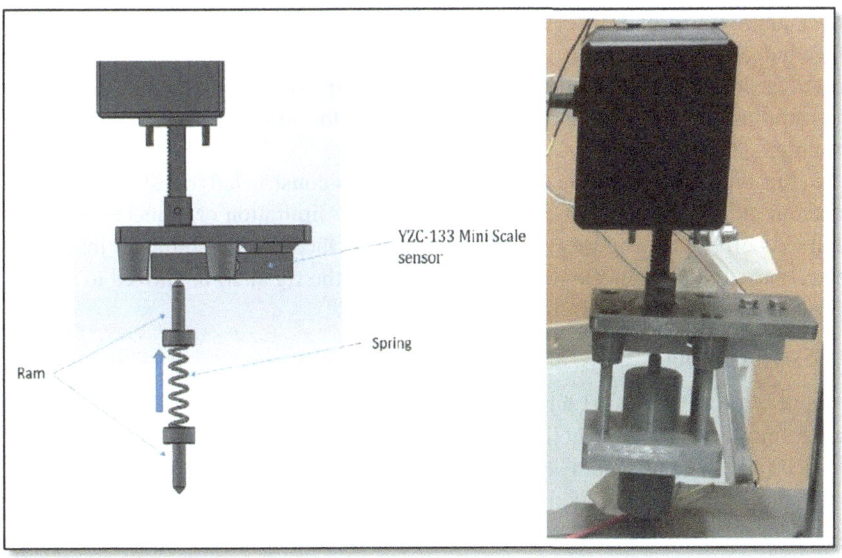

Fig. 7 Endurance rig spring

4.3 Software's Outcome

Concerning the software, it was determined that three types of testing need to be carried out to ensure that the final software is the function in the manner, and it was intended to. There are three types of testings that need to be carried out before the rig can be used to conduct the testing:

- Unit testing
- Integration testing
- System testing.

a. Unit testing

Throughout the first test, the program is submitted to assess specific components to determine whether each one is fully functional and can carry out a given task. The aim of this is to determine if the application and module function as designed, e.g., one of the first test carried out was to determine if an Arduino board can operate an A.R.A linear actuator, this was done by writing a small Arduino sketch that shall retract and extract the motor. Once each component passes its initial unit testing, it could be moved to the next phase of testing which integration is testing.

b. Integration testing

Integration testing gives us the opportunity to combine all supported components within a program and test them as a group. This testing level is intended to discover any defects or limitation between the modules. This is especially valuable since it decides how productively the units are running together. Bearer in mind that regardless of how proficiently every unit is running, on the off chance that they are not appropriately incorporated, it will influence the effectiveness of the software program.

In this type of testing, one of the three units was constructed to test how each unit function individually and to check if there are any limitation or defects that should be fixed before carrying on (Fig. 7, right panel). Once we are confident that one of the constructed units meets the requirements and the rig shall be moved to the next phase of testing which is known as system testing.

c. System testing

System testing is the level where a whole system is tested as one. The objective at this testing is to assess whether the system has conformed to the majority of the defined fundamentals and to see that it satisfies quality guidelines. System testing is vital since it checks that the application meets the specific, practical fundamentals that were set.

5 Conclusions and Future Works

This project started out with a goal of designing and implementing a cost-effective testing rig. Throughout each process, we become closer and closer in having a complete system that shall be used to test the safety system of the chair/shower commode chair. Just like any other project, this project has overcome a majority of its toughest obstacle where the first chosen load cell was deemed not fit for the purpose, and therefore, a new load cell had to be selected which lead to changing some of the rig original designs. All these changes added more time to the project and push the initial time even further than the estimated time that was set for completion.

After extensive discussion, it was concluded that the C# GUI program was no longer needed and therefore only the Arduino sketch needs to be implemented in cooperation with PLX-DAQ software. PLX-DAQ a Microsoft Excel adds on tool that was designed by parallax to acquisition data from microcontroller via serial port and sends the data directly into Excel. This tool provides an easy spreadsheet analysis of data collected from the microcontroller. The final decision to this came down to eliminate the time taken to produce the GUI and spend that time into finalizing the test rig.

At the time being, the rig is on the last huddle of integration testing before moving to system testing; this is due to having no sufficient time to carry out all the tests Once these issues have been fixed; therefore, the two extra A.R.A linear actuator units shall be implemented into the test rig, and the software shall be implemented for the extra units.

This technology may be enriched with further sensorial components [12, 13] in order to prevent injuries. Moreover, this solution may be adopted in other applications especially when the systems involve daily life use and applications for elderly people.

Acknowledgements This work was presented in the dissertation form in fulfillment of the requirements for the degree of computer science and electronic engineering for the student Rashid Abdulrahman under the supervision of Dr. Emanuele Lindo Secco from the Robotics Laboratory, School of Mathematics, Computer Science, and Engineering, Liverpool Hope University.

References

1. K. Schwab, *The Fourth Industrial Revolution* (Portfolio Penguin, London, 2017)
2. R. Allen, *The British Industrial Revolution in Global Perspective* (Cambridge University Press, Cambridge, 2009)
3. World Economic Forum, The Fourth Industrial Revolution, by K. Schwab (n.d.), https://www.weforum.org/about/the-fourth-industrial-revolution-by-klaus-schwab. Accessed 24 Apr 2019
4. G. Khan, D. Isreb, 1, 2, 3… Here comes the 4th Industrial Revolution, Digital Pulse (2018), https://www.digitalpulse.pwc.com.au/fourth-industrial-revolution-guide/?fbclid=IwAR1w3VSADACPnZi7e7IXZv4WeUxSraszRKlElD3NxgI9QTDIyq9I3Rna2As. Accessed 24 Apr 2019

5. R. Reddy, Robotics and intelligent systems in support of society. IEEE Intell. Syst. **21**(3), 24–31 (2006), http://citeseerx.ist.psu.edu/viewdoc/download?doi=10.1.1.86.7405&rep=rep1&type=pdf. Accessed 24 Apr 2019
6. S. Nof, *Handbook of industrial robotics* (Wiley, Hoboken, New Jersey, 2009), pp. 755–773
7. Grandviewresearch.com. Industrial Robotics Market Worth $41.23 Billion By 2020 (2019), https://www.grandviewresearch.com/press-release/global-industrial-robotics-market. Accessed 22 Apr 2019
8. B. Dhillon, *Robot Reliability and Safety* (Springer, New York, New York, 1991)
9. J. Mokyr, *The Enlightened Economy: Britain and the Industrial Revolution, 1700–1850* (Yale University Press, New Haven, 2009)
10. T. Plutschinski, The 4th Industrial Revolution from an Ethical Perspective. ISPSW Strategy Series: Focus on Defense and International Security, (516), 1–5 (2017), http://www.css.ethz.ch/content/dam/ethz/special-interest/gess/cis/center-for-securities-studies/resources/docs/ISPSW-516%20Plutschinski.pdf. Accessed 24 Apr 2019
11. Cortech-healthcare.com. Cortech Healthcare—Innovative products for Independent Living (2019), http://www.cortech-healthcare.com/. Accessed 3 July 2019
12. C. Owen-Bridge, S. Blakeway, E.L. Secco, An integrated & secure system for wearable devices. Special Issue on Multidiscip. Sci. Eng. Adv. Sci. Technol. Eng. Syst. J. **3**(6), 432–437 (2018)
13. E.L. Secco, T.F. Agidew, A.K. Nagar, An optical-based fingertip force sensor. Special Issue on Multidiscip. Sci. Eng. Adv. Sci. Technol. Eng. Syst. J. **3**(5), 23–27 (2018)

Development of an Algorithm for the EMG Control of Prosthetic Hand

Emanuele Lindo Secco, Philippe Caddet, and Atulya K. Nagar

Abstract This work presents the development of a new algorithm for the control of robotic and prosthetic hands: the proposed architecture is made of an EMG wearable sensor and a personalized Graphical User Interface (GUI). The proposed system inherits and processes eight EMG signals which are locally amplified and rectified within the wearable device: then a signal classifier allows piloting a 2 degree of freedom cursor on the GUI in order to reach a provided target in the Cartesian space. The aim of this study is to finally provide a user-friendly interface for training human subjects on reaching movements with the EMG signals of their forearm muscles. This approach as twofold objectives: (1) to maintain, train and support the muscular tone and (2) to provide an interface for the physiotherapy and preparation of prosthetic use in daily life.

Keywords Intelligent systems · EMG · Smart devices · Human–device interaction

1 Introduction

There are more than 1 million annual limb amputations globally, one every 30s [1]. Such amputations are touching daily life of these people since amputee are often affected by huge incapacity to perform basic tasks at their home, work and in their social environment.

E. L. Secco (✉) · P. Caddet · A. K. Nagar
Robotic Laboratory, School of Mathematics, Computer Science and Engineering, Liverpool Hope University, L16 9JD Liverpool, UK
e-mail: seccoe@hope.ac.uk

P. Caddet
e-mail: philippe.caddet@insa-rouen.fr

A. K. Nagar
e-mail: nagara@hope.ac.uk

P. Caddet
Institut National Des Sciences Appliquees (INSA), Rouen, Normandie, France

© Springer Nature Singapore Pte Ltd. 2020
A. K. Nagar et al. (eds.), *Soft Computing for Problem Solving 2019*,
Advances in Intelligent Systems and Computing 1139,
https://doi.org/10.1007/978-981-15-3287-0_15

At the same time, recent scientific developments are providing novel tools in order to help and support such casualties, particularly with the increasing diffusion of robotics and prosthetics [2–4]. An important role on this field is also performed by science in general and the growing of artificial intelligence (AI) which makes some of device in the market smarter and more efficient in terms of higher motion capability, higher biomimetics, better signal interpretation and so on [5, 6].

Here, we are proposing an integrated system where physiological signals can be used to pilot a user-friendly graphical interface in order to enhance the muscular tone of upper limbs while providing a potential tool for the control of prosthetic devices in a "natural" way [7]. The paper is organized into two sections: in the first part—Materials and Methods—we characterize the main components of the system and in the second part—The EMG Graphical User Interface—we focus on the algorithm and outcome of the proposed architecture.

2 Materials and Methods

In this section, the overall system is presented with its own hardware and software components.

2.1 The Processing Environment

Processing is an open-source graphical library and integrated development environment (IDE) or playground "built for the electronic arts, new media art, and visual design communities with the purpose of teaching non-programmers the fundamentals of computer programming in a visual context" [8]. Processing is based on the Java programming language, with the possibility of adding classes and mathematical functions. It allows the development of Graphical User Interface (GUI) and the interaction with input and output of different nature (Fig. 1).

2.2 The Myo Bracelet

The Myo bracelet is a wearable and commercial device for the acquisition, processing and wireless communication of EMG signals, namely the electromyographic output coming from the muscle contraction [9]. The device is made of two main components: the bracelet itself and a USB dongle to be used combined with a computer in order to collect the bracelet information. The device is also provided with software and a set of libraries or SDK for the design of solutions and prototype (Fig. 2).

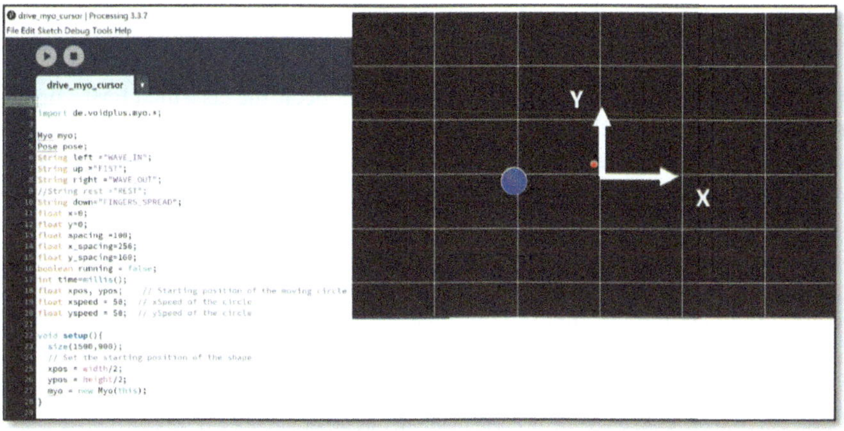

Fig. 1 An overview of the proposed algorithm in Processing programming language (left panel) and its resulting graphical outcome (right panel)

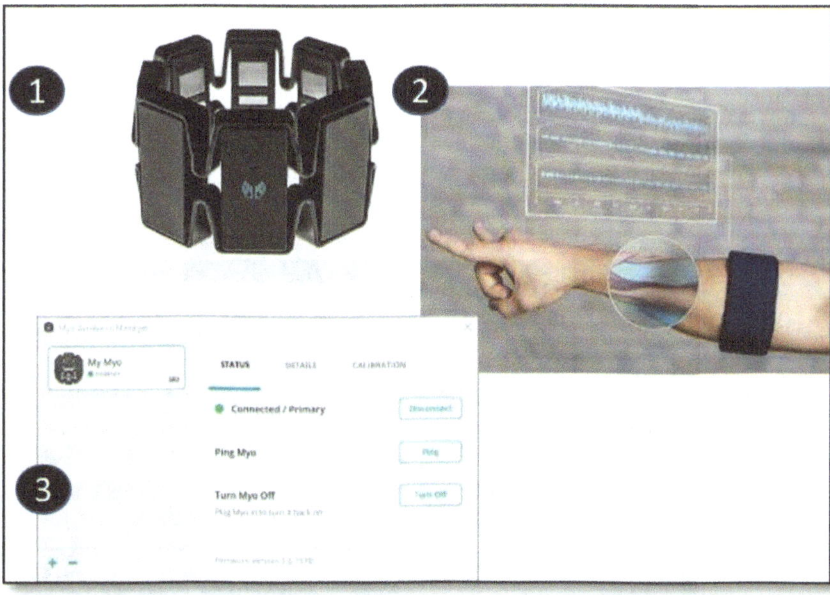

Fig. 2 The Myo bracelet (1), a pictorial representation of its functioning (2) and the default user interface software mask (3) (*Source* RobotShop and Thalamic Lab)

3 The EMG Graphical User Interface

An EMG GUI has been developed: the interface collects the classified EMG signals from the wearable bracelet and allows the navigation of a mobile cursor on a grid which is projected on the screen. Namely, the moving cursor is activated and driven by the Myo signals, whereas a static cursor, i.e., the target, is provided through the input of x, y Cartesians coordinates. This navigation cursor aims at reaching a random target on the grid (Fig. 3).

As soon as the target is reached, the navigation cursor is returned to the home position (namely the position of coordinates $x = 0$ and $y = 0$) and a novel target is provided. The interface is managed through the following structured sequence of events which are implemented in the controlling algorithm under the Processing programming language:

- Step 1—Target positions (x_{tg}, y_{tg}) are acquired from a list of target provided via an Excel file
- Step 2—EMG signals are real-time detect from the forearm muscles' contraction
- Step 3—An embedded classification of the EMG signals is performed in order to identify four different hand configurations, namely (Fig. 4):
  - *Wave in*
  - *Wave out*
  - *Fist*
  - *Open Hand*

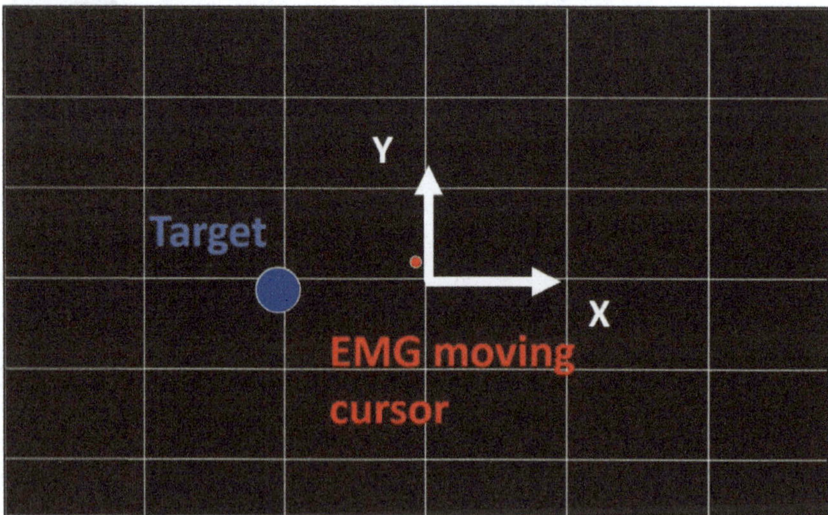

Fig. 3 Overview of the EMG Graphical User Interface with the target or static marker and the EMG cursor or mobile marker, in blue and red color, respectively

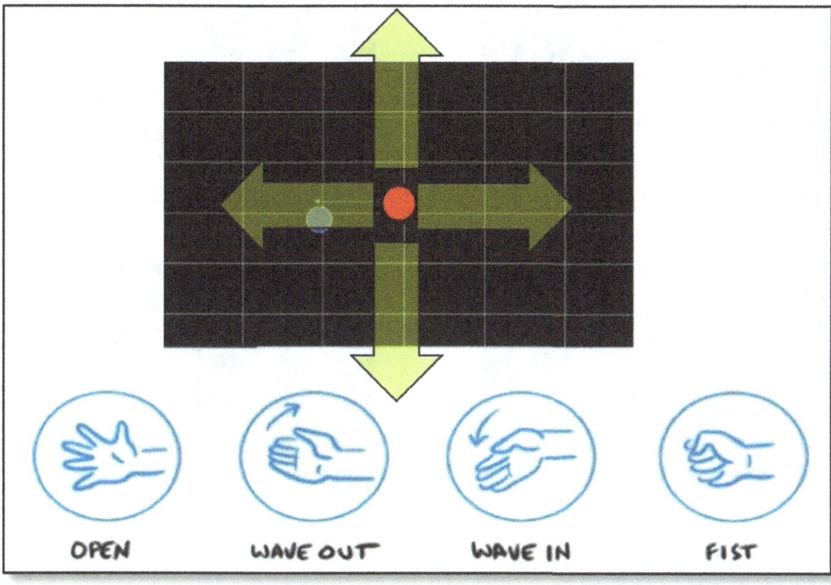

Fig. 4 Overview of the EMG classification (as internally performed within the Myo/Thalamic Software) and of the four directions of the EMG mobile cursor

- Step 4—A conversion of the assigned class is performed into four different possible movements of the mobile cursor within the projected grid, according to the following mapping (Fig. 5):
  - *Wave in* maps into a cursor movement to the *left* of the grid
  - *Wave out* maps into a cursor movement to the *right* of the grid
  - *Fist* maps into a cursor movement to the *top* of the grid
  - *Open Hand* maps into a cursor movement to the *bottom* of the grid
- Step 5—The gird is updated with the projection of the novel cursor position vs the target reference
- Step 6—If the target is not reached, the process proceeds repeatedly from step 2
- Step 7—Otherwise the process terminates when the target is reached.

4 Preliminary Tests

A set of preliminary tests of the interface has been performed: during these tests, the x_{tg} and y_{tg} target positions were available through an Excel table as mentioned in Sect. 3.

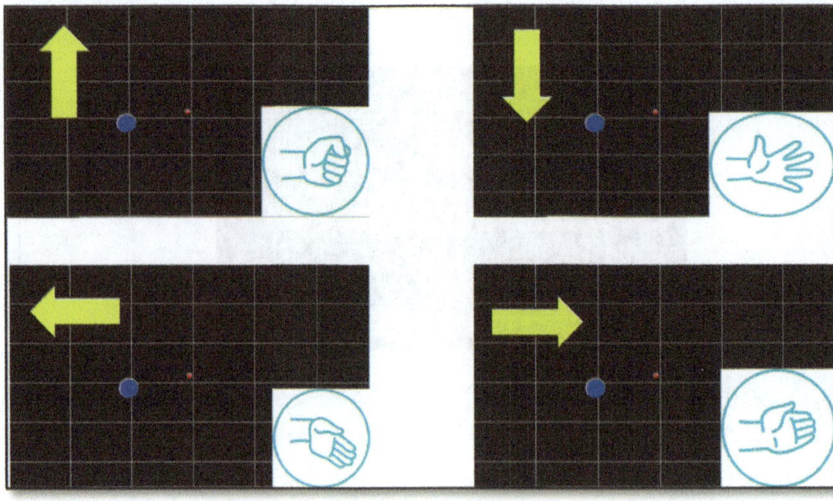

Fig. 5 The mapping occurring between the EMG classification (i.e., the muscle contraction of the subject) and the movement of the mobile cursor on the GUI

Meanwhile, the time required to perform the cursor movement was also measured. An example of the data structured reporting the time (in ms) is reported in Fig. 6.

These tests aimed at checking the functionality of the system, even if they only involved the software components. Further, tests should validate the feasibility of having human subjects able to improve their capability on moving the cursor into the target as soon as different consequent targets are given along the horizontal and vertical axes of the grid.

5 Conclusion

We presented a novel interface based on a new algorithm which implements a set of Cartesian movements on the screen of a computer, namely the grid. Such movements are designed in order to be induced by the muscle contraction of human subjects while trying to reach and match provided target within the grid.

Preliminary tests were performed on the software components of the project: a set of proper tests should be developed where the time required to reach such targets should be measured vs extensive protocols where the subjects improve their performance trial by trial. This setup should help and support a variety of applications [10, 11], from the control of prosthetic devices to the improvement of the muscular tone of impaired people in a rehabilitative context [12, 13].

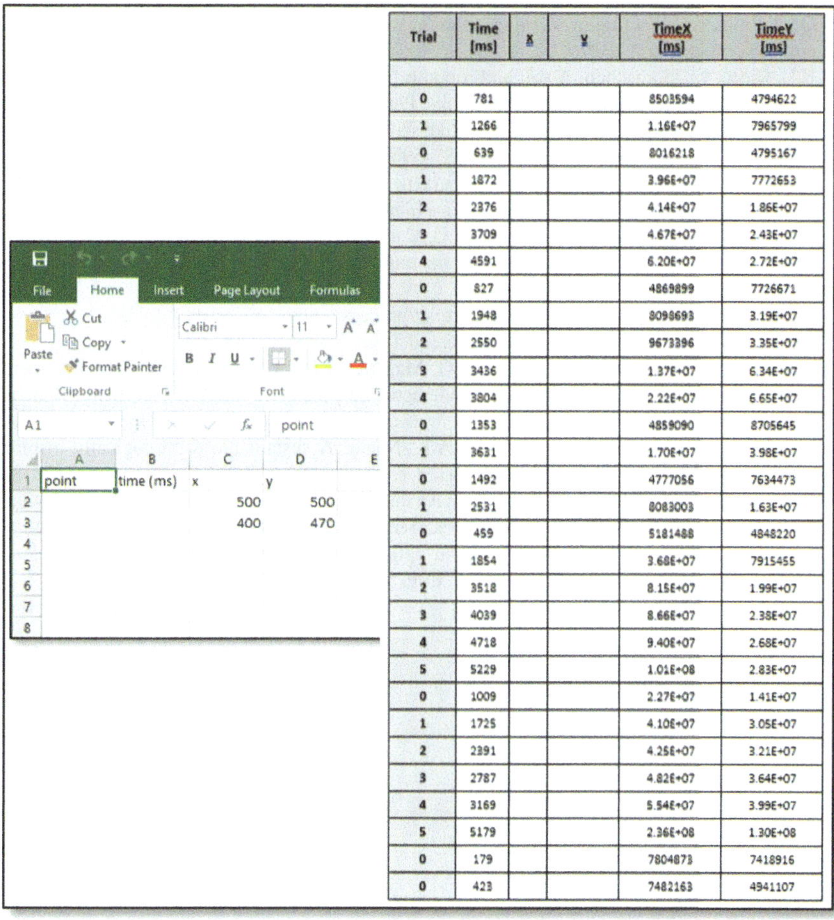

Trial	Time [ms]	x	y	TimeX [ms]	TimeY [ms]
0	781			8503594	4794622
1	1266			1.16E+07	7965799
0	639			8016218	4795167
1	1872			3.96E+07	7772653
2	2376			4.14E+07	1.86E+07
3	3709			4.67E+07	2.43E+07
4	4591			6.20E+07	2.72E+07
0	827			4869899	7726671
1	1948			8098693	3.19E+07
2	2550			9673396	3.35E+07
3	3436			1.37E+07	6.34E+07
4	3804			2.22E+07	6.65E+07
0	1353			4859090	8705645
1	3631			1.70E+07	3.98E+07
0	1492			4777056	7634473
1	2531			8083003	1.63E+07
0	459			5181488	4848220
1	1854			3.68E+07	7915455
2	3518			8.15E+07	1.99E+07
3	4039			8.66E+07	2.38E+07
4	4718			9.40E+07	2.68E+07
5	5229			1.01E+08	2.83E+07
0	1009			2.27E+07	1.41E+07
1	1725			4.10E+07	3.05E+07
2	2391			4.25E+07	3.21E+07
3	2787			4.82E+07	3.64E+07
4	3169			5.54E+07	3.99E+07
5	5179			2.36E+08	1.30E+08
0	179			7804873	7418916
0	423			7482163	4941107

Fig. 5 Results of the preliminary tests as performed with the GUI

Acknowledgements This work was presented in report form in fulfillment of the requirements for the Internship for the student Philippe Cadett under the supervision of Dr. Emanuele Lindo Secco from the Robotics Laboratory, School of Mathematics, Computer Science and Engineering, Liverpool Hope University.

References

1. Access Prosthetics, 15 Limb loss statistics that may surprise you—access prosthetics (2019), https://accessprosthetics.com/15-limb-loss-statistics-may-surprise/. Accessed 4 Jul 2019
2. Touchbionics.com, Homel touch bionics (2019), https://www.touchbionics.com/. Accessed 4 Jul 2019

3. Open Bionics, *Open bionics—turning disabilities into superpowers* (2019), https://openbionics.com/. Accessed 4 Jul 2019
4. E.L. Secco, C. Moutschen, A soft anthropomorphic & tactile fingertip for low-cost prosthetic & robotic applications. EAI Endorsed Trans. Pervasive Health Technol. 4(14)
5. G. Magenes, S. Ramat, E.L. Secco, Life-like sensorimotor control: from biological systems to artifacts. Curr. Psychol. Cogn. **21**(4/5), 565–596 (2002)
6. E.L. Secco, G. Magenes, A feedforward neural network controlling a 3 dof finger, in *Proceedings Fifth International Conference on Cognitive and Neural Systems*, Boston, MA (2001)
7. G. Matrone, C. Cipriani, E.L. Secco, M.C. Carrozza, G. Magenes, A biomimetic approach based on principal components analysis for multi-DoF prosthetic hand control, in *Workshop CORNER Genova–IIT*, 14–15 (2009)
8. Processing.org, Processing.org (2019), https://processing.org. Accessed 4 Jul 2019
9. Myo (2017), https://www.myo.com. Accessed 4 Jan 2017
10. G.C. Matrone, C. Cipriani, E.L. Secco, G. Magenes, M.C. Carrozza, Principal components analysis based control of a multi-dof underactuated prosthetic hand. J. Neuroeng. Rehabil. **7**(1), 16 (2010)
11. E.L. Secco, C. Moutschen, T. Agidew, A.K. Nagar, A sustainable & biologically inspired prosthetic hand for healthcare. EAI Endorsed Trans. Pervasive Health Technol. 4(14) (2018)
12. C. Owen-Bridge, S. Blakeway, E.L. Secco, An integrated & secure system for wearable devices. Spec. Issue Multidiscip. Sci. Eng. Adv. Sci., Technol. Eng. Syst. J. **3**(6), 432–437 (2018)
13. E.L. Secco, T.F. Agidew, A.K. Nagar, An optical-based fingertip force sensor. Spec. Issue Multidiscip. Sci. Eng. Adv. Sci., Technol. Eng. Syst. J. **3**(5), 23–27 (2018)

Improving Data Quality in the Cargo Industry with Modern Recurrent Neural Network Architecture

Lewis Wong, Declan O'Connor, Neil Buckley, and Atulya K. Nagar

Abstract The shipping industry is a multifaceted trading system, enabling the flow of goods between countries. Often the standards, methods and technologies become interspersed and difficult to plan, disrupting the supply chain. Data is not always accurate, necessitating further analysis. This paper presents an approach to solving key issues in two important facets of the supply chain, predicting the date of restitution for a cargo container and using an optical character recognition (OCR)-centred pipeline to extrapolate data from containers. Both approaches use long short-term memory (LSTM) models, including bi-directional LSTMs. These methods leverage state-of-the-art text recognition architecture and advanced algorithm ensembling, giving significant improvements in data quality. The experimental results illustrate that these methods vastly outperform current industry practices and more recent approaches.

Keywords Time series prediction · Optical character recognition · Data quality · Supply chain · Shipping

1 Introduction

Transport is a growing industry. The European Commission for Transport states in *Mobility and Transport* [1] that by 2050 the freight transport industry will increase by 60%. Of all modes of transport freight and road hold, this has the largest share of

---

L. Wong · D. O'Connor · N. Buckley (✉) · A. K. Nagar
School of Mathematics, Computer Science and Engineering, Liverpool Hope University, Liverpool, UK
e-mail: bucklen@hope.ac.uk

L. Wong
e-mail: 14003777@hope.ac.uk

D. O'Connor
e-mail: 13007972@hope.ac.uk

A. K. Nagar
e-mail: Atulya.Nagar@hope.ac.uk

© Springer Nature Singapore Pte Ltd. 2020

A. K. Nagar et al. (eds.), *Soft Computing for Problem Solving 2019*,
Advances in Intelligent Systems and Computing 1139,
https://doi.org/10.1007/978-981-15-3287-0_16

activity, with around 54%, as it dominates over short to medium distances. Optimal planning is non-trivial, being a complex temporal problem with nonlinear relationships, which makes finding an explicit formula for accurate prediction an ambitious task. Currently, only part of the event cycle has advanced methods for tracking. Containers do not currently have global positioning system (GPS) devices attached, so real-time data cannot be obtained as they can for ships, which use an automatic identification system (AIS). Companies such as Marine Traffic provide high-quality location data and trajectories internationally. Once a container is on land, there can be gaps between milestone events being recorded. Additionally, there is no universal system to predict when the container will be unloaded, gated out, delivered and returned to the port empty, i.e. restituted.

Warehouses are responsible for updating journey data by hand, which often leads to data being missing or containing human errors. Our proposed long short-term memory (LSTM) solution can automate this process by reading container technical information and attach a time stamp to it. The Bureau International des Containers et du transport (BIC) maintains a register of containers and their technical details. The BIC BoxTech database allows for the validation of containers upon delivery, increasing supply chain visibility and aiding future planning, while also ensuring all physical containers are documented. By using the proposed method with BIC's database, we can automatically provide the supply chain partners with the most recent container technical information.

Supply chains have as many as 30 actors involved across multiple international borders. Often the planned journey is not representative of the actual journey. Retailers and automotive companies try to keep minimal stock levels, preferring to use a 'just in time' methodology. This is because supply chains can incur high costs when they do not perform as planned. Projections of key milestone dates would also benefit the environment, and estimated delivery and restitution dates would allow consignees to prevent costs from accumulating as they can often run into the millions per annum.

This paper proposes LSTM, a deep neural recurrent model, along with LSTM-based OCR, to achieve the following objectives:

1. The projection of the restitution date to allow importers to have early warning of the potential of costs and ensure data is provided at the end of the journey.
2. An automated process for importers to capture necessary regulatory container information and provide a delivery date that can be validated automatically.

The remainder of this paper is structured as follows: Sect. 2 reviews prior literature, Sect. 3 details the use of LSTM for container restitution prediction and automated container detail reading, Sect. 4 discusses the results and carries out a comparative analysis, and Sect. 5 concludes the study with a detailed analysis of future enhancements.

2 Related Work

Bodunov et al. [2] demonstrate a solution for predicting the arrival of Maritime Traffic using geo-spatial data. The approach used an ensemble method of gradient boosted decision trees such as XGBoost and ERT. The paper also discusses an attempt to implement LSTM but leaves it for future work.

Similarly, Wang et al. [3] proposed a 'wide' and 'deep' neural network that can account for feature interactions via a cross product. Using a combination of spatial, temporal and traffic data with personalised and augmented information. Wang created a network structure that could utilise modern machine learning techniques for the numerical, categorical and sequential elements separately before recombining them for the output.

Fernando et al. [4] proposed an OCR system to identify container ownership codes. The aim was to help crane operators identify which container they are handling. They used Tesseract with image processing. An image was taken from the crane of the container then processed and cropped before being fed into Tesseract. They also used a whitelist to ensure the correct characters where identified.

The work of [5], similar to the work in this paper, proposed an OCR engine that utilised a convolutional neural network (CNN) to extract container codes. They used multiple images of the same container to cross-reference results. Two LeNet architectures were used, one to recognise the alphabet (A–Z) and another for digits (0–9).

A deep neural network architecture was also used by Yi et al. [6], who proposed an OCR system based on machine learning. The method used two modules, one for detection and the other for recognition. They combined two networks to make up for each other's insufficiencies. Similar to our work, they implemented a fail-safe technique to ensure the correct amount of digits/characters was returned. This ultimately reduced errors in the data.

3 Method

LSTM neurons are crafted to handle long-term dependencies. There have been many variations on how LSTM neurons function, yet the defining characteristic is having three gates: an input gate (it), output gate (ot) and forget gate (ft). This is shown in Fig. 1.

Most important is the forget gate, which performs the following calculation.

$$f_t = \sigma(W_f \cdot [H_{t-1}, x_t] + b_f) \tag{1}$$

This decides what to remove from the current cell state. It takes the values of the current layer x^t and the previous iteration H_{t-1} and outputs a value between zero and one for every cell in the context cell C_{t-1}. A zero will cancel the input while

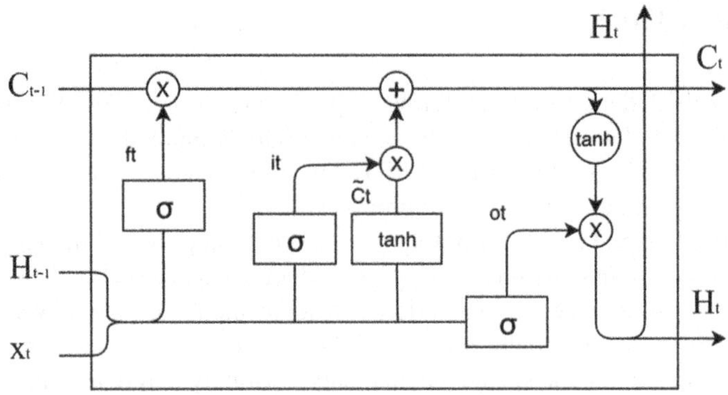

Fig. 1 LSTM neuron architecture

the one will allow the data to persist. This context state gives the network a frame of reference, for example, it could 'remember' a port having delays due to increased regulatory container checks for various unforeseen reasons. In the case of OCR, it can remember segmented letter sequences to formulate words.

The purpose of the following task is calculate what to retain. This is in two parts, given in (2 and 3). The first is a sigmoid layer, known as the input gate and denoted as i_t. This decides what is to be added. There is also a tanh layer, $\tilde{C}$, which creates new candidate values that can be added. These two gates are combined by a multiplication pointwise operation and added to the context cell updating the network.

$$i_t = \sigma(W_i \cdot [H_{t-1}, x_t] + b_i) \tag{2}$$

$$\tilde{C}_t = \tanh(W_C \cdot [H_{t-1}, x_t] + b_C) \tag{3}$$

Once the context neuron has been updated, it can then be used to update the next cell's context neuron. The forget gate and the input gate previously discussed are combined to give this output.

$$C_t = f_t * C_{t-1} + i_t * \tilde{C}_t \tag{4}$$

The neuron finally needs to decide what to output to the next layer, a sigma function decides the cell state. The context cell modified by a tanh function is multiplied by the sigma output giving the final output, as in (5) and (6). For this time-based task in this research, this will give relevant information for the future, for example, if a company is making an effort to resituate containers more quickly, and this is observed in the model, this may be indicative of their future containers being resituated in a more timely manner. Simply put, this is a long-term dependency that can be mapped.

$$o_t = \sigma(W_o[h_{t-1}, x_t] + b_o) \tag{5}$$

$$H_t = o_t * \tanh(C_t) \tag{6}$$

The data used for time series analysis in this research is a compilation of data sourced from carriers, consignees and suppliers. It is largely dependent on five key time stamps that operate in sequential order. Each journey has the following events denoted by the arrows in Fig. 2, with each event corresponding to a time stamp. Location events are denoted by rectangles, with the three grouped ones at the top referring to the port operations and the bottom one being external facility operations that pertain to the consignee.

Data is stored in a tabular format, with each row signifying the journey of one container. Categorical data, such as carrier name and port name along with other variables have been encoded for additional predictive power. Feature engineering allowed for the extrapolation of additional date-time features, such as week, month and year. Time has been preserved by calculating the difference between each timestamp and using this integer value as a new time feature.

With machine learning techniques at the forefront of travel time estimation [7], LSTMs are adept at dealing with time-based problems, due to capturing long-term

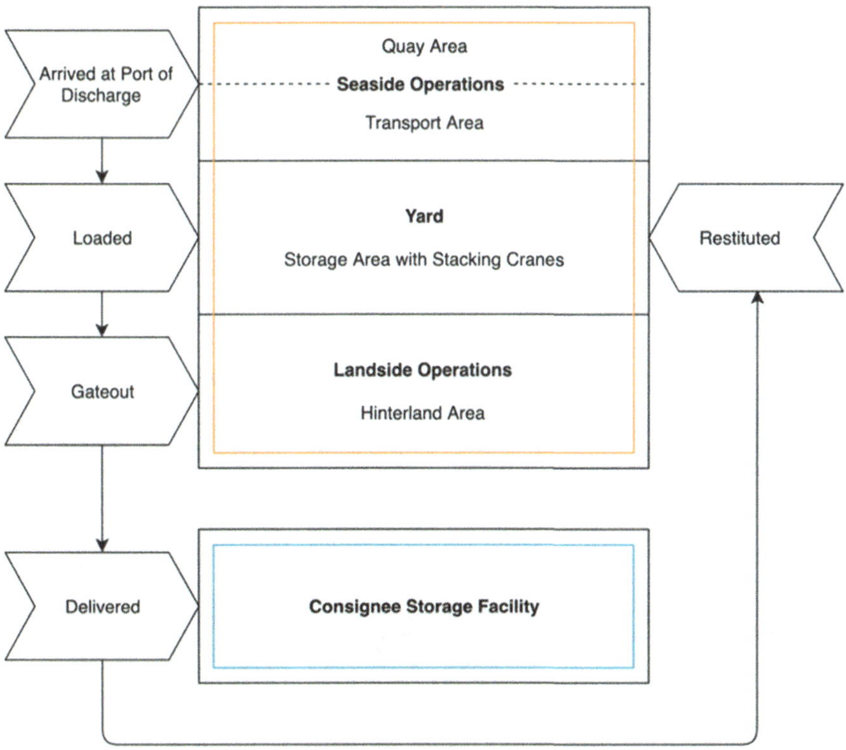

Fig. 2 Container terminal and importer facility

dependencies. For the proposed solution, a bi-directional variant is used for the two-way sequential data. The model uses Nadam optimisation and ReLu activation, with a four-layered structure including dropout to prevent overfitting. To enable the use of an LSTM, the journey data was encoded and reshaped into a 3D tensor.

The model was trained for 80 epochs with a batch size of 128, and then leveraged with two gradient boosted algorithms: XGBoost and CatBoost. The gradient boosted algorithms were better at handling categorical variables and one-hot-encoded extracted time features, while the LSTM model is better at capturing temporal sequences using the data captured as the difference between time steps. The ensemble is a weighted summation of the predicted values for each journey. It uses three unique models trained on the sample size of 14664 journeys, validated using a sample of 4889 and tested with a sample of the same size. The measured results of this ensemble performed better than the sum of its parts.

To accompany that model, the LSTM-based OCR system is at a proof of concept stage, where the focus is to accept container images from multiple sources such as (a) port gate camera feeds (b) loading photos including container door images and convert the text into a machine-readable format. To do this, the region of interest (ROI) is cropped. This is the right side of the container door where the technical information is located.

The image is then ready to be processed. It is converted to grayscale and binarised with a global thresholding algorithm. A dilation filter is applied to increase the size of the white regions. A contour algorithm is then applied, joining all continuous points (along a boundary) that have the same colour. This identifies the white text areas in the image. Finally, all the contours (text areas) are cropped from the original and sent for further processing, before being sent to Tesseract, an open-source LSTM-based OCR engine (Fig. 3).

Each cropped ROI is processed in the following order: (1) Grayscale, (2) enlarge image, (3) remove blurring, (4) noise removal, (5) invert colours and (6) thresholding algorithm. Black text with a white background is what remains in each image. They

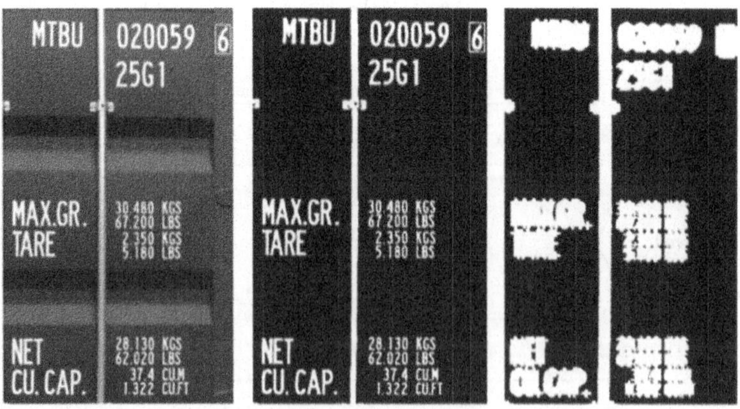

Fig. 3 Cropped ROI: grayscale (left) binarised (middle) dilation filter (right)

Fig. 4 Cropped ROI: grayscale (left) remove blurring (left middle) invert colours (right middle) thresholding (right)

are then fed into Tesseract stochastically. A string is returned for each image, and regular expressions are used to match the returned strings to the corresponding six components on a container (Fig. 4).

1. Owner code: String of four letters. The last letter is always either J, U, R or Z.
2. Container code: Number containing six digits.
3. Check digit: Single digit.
4. Type code: String with length of four, consisting of numbers and letters.
5. Containers weights: Total of three weights in both LBS and KGS. Max gross and net weight are five digits long. Tare weight is four digits long.
6. Cubic capacity: Two numbers. The length in metres is three digits. The length in feet is four digits.

The only three-digit number on a container is the volume in cubic metres, so it can be matched straight away. This leaves the remaining with four-digit and five-digit numbers, which can be put into descending order. From here, we can determine which weight the number corresponds to.

The results are validated by checking the owner code and type codes against a list of valid codes. A final check is carried out to identify any missing data. If a weight was not recognised, e.g., if the tare KGS is missing but the LBS present, it is converted to provide a fuller dataset. The process of filling in any missing data is as follows:

1. Check which data points are missing.
2. Identify if at least one of the KGS or LBS data points have been identified.
3. Convert one data point to the other.

Once the data validation has been completed, the remaining data is converted into JSON format.

4 Discussion and Results

The results in Table 1 demonstrate the capability of the proposed method. It turns out that the model is able to predict a restitution date vastly outperforming current industry practices. Conceptually, the model was designed to use LSTM in a similar way to that of a classical forecasting techniques, while tree-based methods take advantage of the extrapolated time features and exogenous categorical variables.

Table 1 Comparative analysis of models with lift chart

	Bi-directional LSTM	CatBoost	XGBoost	Ensemble method
Three time values	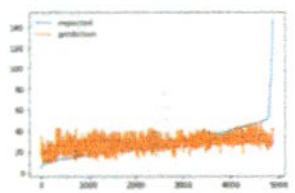	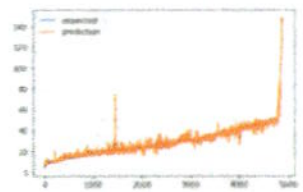	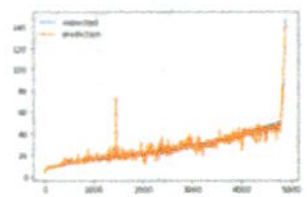	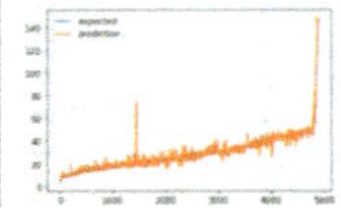
Two time values	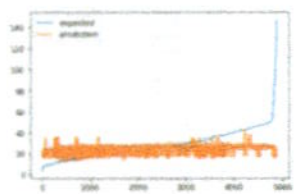	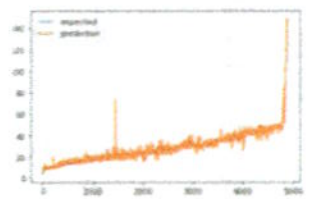	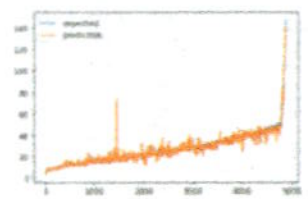	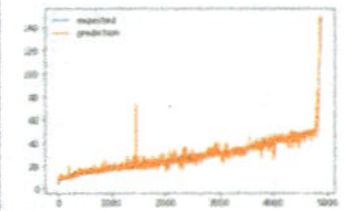
One time values	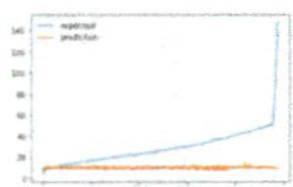	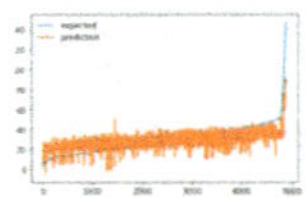	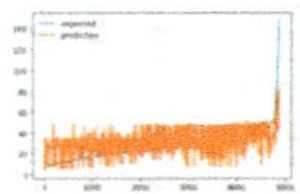	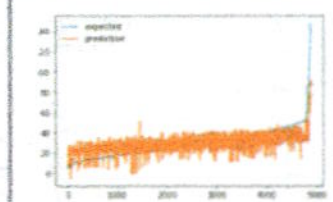
No time values	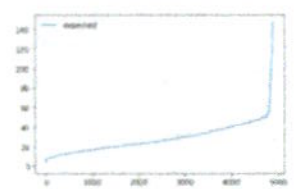	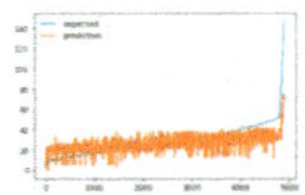	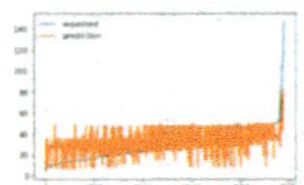	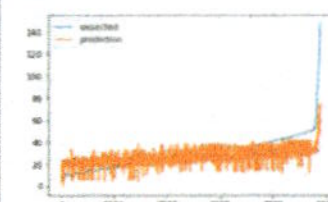

Table 2 Comparative analysis with root-mean-square-deviation (RMSE) values

	Bi-directional LSTM	CatBoost	XGBoost	Ensemble method
Three time values	13.44380	2.27831	2.26504	2.23366
Two time values	17.40835	3.48588	2.46461	2.34111
One time values	25.43657	12.22145	13.33857	10.59285
No time values	N/A	14.190778	13.65147	11.40971

Table 2 shows the RMSE metric results from the model in the same format as Table 1. It demonstrates the accuracy measured in days.

$$\mathrm{RMSE} = \sqrt{\frac{1}{N}\sum_{i=1}^{N}\left(y_i - \hat{y}_i\right)^2}. \tag{14}$$

RMSE is the square root of the ground truth minus the prediction squared for the sum of all predictions.

The results observed are significant enough for consignees to be able to track when their containers should be restituted, preventing being charged a fee. It also allows importers the ability to catch early containers that are expected to take longer to restitute and prioritise those being returned. In turn, this will reduce overall costs to those who are importing and allow ports to operate with higher and managers to make clearer economic decisions.

For the OCR side of the solution, the four most commonly used colours for containers where tested. The result showed that it is possible to collect enough data from a container image to enable the automation of the VGM process. This also allowed for the delivery milestone to be updated and verified, resulting in more accurate restitution date prediction. It also fared well against the leading techniques and allowed for more data to be collected overall.

This improves data quality by enhancing global container databases to have the most up-to-date container information, enhancing supply chain visibility. The system decreased the likelihood or errors occurring from the human labelling process.

Table 3 shows that the proposed method is able to compete with some of the leading industry techniques. Classical image processing techniques used with an

Table 3 Comparison of OCR models

Model	Images tested	Accuracy (%)
CNN	N/A	95.41
CTPN & MSER	N/A	93
SVM	9260	92.60
ART2	79	98.5
Our method	300	91.3

Table 4 Average data points recognised and converted from tesseract

Container colour	Data points recognised by tesseract out of 12	Data points calculated	Total data points returned out of 12
Brown	6	4	10
Blue	10	2	12
Green	10	2	12
Red	8	2	10

Note The results are only comparative of the container code of the container technical information. Our method yields ten times more data collected

LSTM-based model has a slightly lower accuracy rating in comparison with others but the overall data collected is far greater.

The results in Table 4 give an average score from the 300 images tested. The average of all the data points from the different container colours was taken to show the robustness of the system. The proposed conversion technique allowed for more data to be returned even when the OCR does not recognise the data point.

5 Conclusion and On-Going Research

The results of this research prove that tree-based methods and recurrent neural networks can capture highly complex nonlinear relationships between variables and trip duration. LSTM-based OCR used in with classical image processing techniques can capture the full technical details from container doors.

Both methods provide much more accurate data to those who need it most in the supply chain. The best results for the restitution prediction came from an ensemble-based method that utilised two different machine learning models with the LSTM model. The highest accuracy results from the OCR system came from container images that were successfully processed.

The time series approach may not be efficient in terms of computational cost but performs exceptionally well in comparison with current industry methods. Similarly, the OCR model may have lower accuracy in comparison with leading techniques but captured a significantly larger portion of the container details. By proving the restitution date, the consignees can track when their containers should be restituted to prevent a demurrage fee in a number of ways.

The future of time-based research and OCR systems lies in attempting to improve the quality of data in the field, both methods create a positive feedback loop that improves data over time and would allow for future results to be more accurate. More sophisticated methods could be employed to fill gaps in the data such as training a network to be able to fill in null values.

Beyond data quality, the ensemble method could benefit from incorporating forecasting techniques, yet the data needs to be broken down into subcategories, such as port, to reduce noise. Furthermore, better quality images would benefit the accuracy of capturing all the container data and by utilising fail-safe data validation techniques, a valid fuller dataset can be provided.

In further research, clustering will be used to determine how the data should be decomposed further. A stacked generalisation method will be employed instead of a basic ensemble method. This is the practice of building a neural network to pick the best ensemble based on the output. Similarly, an adaptive thresholding algorithm will be used on images to reduce noise effecting the results.

The method proposed in this work will reduce costs to all parties, optimise the supply chain and help eliminate manual data entry. It will most importantly improve data quality to not only increase supply chain visibility and efficiency but inspire future machine learning usage in this challenging industry [8].

References

1. Artificial Intelligence in Transport, European Commission (2019)
2. O. Bodunov, F. Schmidt, A. Martin, A. Brito, C. Fetzer, Grand challenge: real-time destination and ETA prediction for maritime traffic (2018). arXiv:1810.05567
3. Z. Wang, K. Fu, J. Ye, Learning to estimate the travel time. In *Proceedings of the 24th ACM SIGKDD International Conference on Knowledge Discovery & Data Mining* (ACM, 2018), pp. 858–866
4. F. Martín Rodríguez, X. Fernández Hermida, J.C. Torres Barragáns, Automatic reading of intermodal containers ID numbers (ownership codes). Instrum. Viewp. **18**, 37–38 (2015)
5. C. Roeksukrungrueang, T. Kusonthammrat, N. Kunapronsujarit, T.N. Aruwong, S. Chivapreecha, An implementation of automatic container number recognition system. In *2018 International Workshop on Advanced Image Technology (IWAIT)* (IEEE, 2018), pp. 1–4
6. Y. Liu, T. Li, L. Jiang, X. Liang, Container-code recognition system based on computer vision and deep neural networks. In *AIP Conference Proceedings*, vol. 1955, No. 1 (AIP Publishing, 2018), p. 040118
7. T. G. Dietterich, Machine learning for sequential data: a review. In *Joint IAPR International Workshops on Statistical Techniques in Pattern Recognition (SPR) and Structural and Syntactic Pattern Recognition (SSPR)* (Springer, Berlin, 2002)
8. Mobility and Transport, European Commission (2019)

Temporal Convolution in Spiking Neural Networks: A Bio-mimetic Paradigm

David Reid and Emanuele Lindo Secco

Abstract Recent spectacular advances in *artificial intelligence* (AI), in large, be attributed to the developments in *deep learning* (DL). In essence, DL is not a new concept. In many respects, DL shares characteristics of "traditional" types of *neural network* (NN). The main distinguishing feature is that it uses many more layers in order to learn increasingly complex features. Each layer convolutes into the previous by simplifying and applying a function upon a subsection of that layer. Deep learning's fantastic success can be attributed to dedicated researchers experimenting with many different groundbreaking techniques, but also some of its triumphs can also be attributed to fortune. It was the right technique at the right time. To function effectively, DL mainly requires two things: (a) vast amounts of training data and (b) a very specific type of computational capacity. These two respective requirements have been amply met with the growth of the Internet and the rapid development of GPUs. As such DL is an almost perfect fit for today's technologies. However, DL is only a very rough approximation of how the brain works. More recently, *spiking neural networks* (SNNs) have tried to simulate biological phenomena in a more realistic way. In SNNs, information is transmitted as discreet spikes of data rather than a continuous weight or a differentiable activation function. In practical terms, this means that far more nuanced interactions can occur between neurons and that the network can run far more efficiently (e.g., in terms of calculations needed and therefore overall power requirements). Nevertheless, the big problem with SNNs is that unlike DL it does not "fit" well with existing technologies. Worst still is that no one has yet come up with definitive way to make SNNs function at a "deep" level. The difficulty is that in essence "*deep*" and "*spiking*" refer to fundamentally different characteristics of a neural network: "*spiking*" focuses on the activation of individual neurons, whereas "*deep*" concerns itself to the network architecture itself

D. Reid (✉)
Artificial Intelligence Lab, School of Mathematics, Computer Science and Engineering, Liverpool Hope University, Hope Park, Liverpool L16 9JD, UK
e-mail: reidd@hope.ac.ul

D. Reid · E. L. Secco
Robotic Lab, School of Mathematics, Computer Science and Engineering, Liverpool Hope University, Hope Park, Liverpool L16 9JD, UK

© Springer Nature Singapore Pte Ltd. 2020

A. K. Nagar et al. (eds.), *Soft Computing for Problem Solving 2019*,
Advances in Intelligent Systems and Computing 1139,
https://doi.org/10.1007/978-981-15-3287-0_17

Pfeiffer and Pfeil (Front Neurosci 12, 2018) [1]. However, these two methods are in fact not contradictory, but have so far been developed in isolation from each other due to the prevailing technology driving each technique and the fundamental conceptual distance between each of the two biological paradigms. If advances in AI are to continue at the present rate, then new technologies are going to be developed and the contradictory aspects of DL and SNN are going to have to be reconciled. Very recently, there have been a handful of attempts to amalgamate DL and SNN in a variety of ways Tavanaei et al. (Neural Netw 111:47–63, 2019) [2] one of the most exciting being the creation of a specific hierarchical learning paradigm in r*ecurrent SNN* (RSNNs) called e-prop Bellec et al. (bioRxiv, 2019) [3]. However, this paper posits that this has been made problematic because a fundamental agent in the way the biological brain functions has been missing from each paradigm, and that if this is included in a new model, then the union between DL and RSNN can be made in a more *harmonious manner*. The missing piece to the jigsaw, in fact, is the glial cell, and the unacknowledged function it plays in neural processing. In this context, this paper examines how DL and SNN can be combined, and how glial dynamics cannot only address outstanding issues with the existing individual paradigms—for example, the "weight transport" problem but also act as the "glue"—e.g., pun intended—between these two paradigms. This idea has a direct parallel with the idea of convolution in DL but has the added dimension of time. It is important not only *where* events happen but also *when* events occur in this new paradigm. The synergy between these two powerful paradigms gives hints at the direction and potential of what could be an important part of the next wave of development in AI.

1 Deep Learning and Recurrent Spiking Neural Networks

An essential distinction between *deep learning* (DL) and *recurrent spiking neural networks* (RSNNs) is in the way the information flows from one neuron to another. DL can be seen as continuous function approximators in which the information is *synchronously* processed, whereas RSNN can be seen as transporting discrete packets of information that can be processed *asynchronously*. DL cares nothing about precisely when something happens, whereas a key characteristic of RSNN is the encoding of temporal information. At a fundamental level, it would not be an unreasonable to consider the difference between DL and SNNs as being equivalent to the distinction between analog and digital neural networks.

Recently, there has been growing interest in applying some of the techniques used in DL to the field of RSNNs [4]. This is undoubtedly because of DL popularity and its success in solving problems that where intractable even a few years ago. As noted in the abstract, the dominance of DL can be attributed to be the right fit for the current technology, but it is also because it can utilize larger-scale cognitive structures, whereas RSNN research has very often only focused on local interaction.

Early attempts to marry DL and RSNN involved transposing either the activation functions or weights in a deep learning system into a firing rate at an equivalent level in a RSNN. However, because RSNNs can use the timing between spikes as information [5], or the absolute timing of an event [6], they can model far more complex processes than deep learning. This allows for far more rapid decision-making in the RSNN system as the synchronous calculation required by deep learning is not required. Taking this idea further, RSNN research opens up the idea of "temporal codes" that may trigger different reactions in different subtle circumstances [7, 8]. Using current technologies, this will come at the cost of longer training times, because fully spiking simulations are computationally more expensive than simulations of traditional neural networks. However, the number of spikes needed is typically far lower than in rate-coded simulations of SNNs, and this opens up the opportunity for both far more efficient and simulation of more complex behavior than is possible with DL.

Despite this supervised training methods in SNN have not fundamentally changed, they were adapted from traditional neural networks. *Spike time-dependent plasticity* (STDP) is the most popular variant of the Hebbian rule for learning. STDP which rewards or punishes the firing of related neurons based on the time of the firing. If a neuron fires soon after another neuron fires, it is heavily rewarded; if it preempts the firing pattern, it is punished. In this way, inputs that might be the cause of the post-synaptic neuron's excitation are made even more likely to contribute to the network, whereas inputs that are not the cause of the postsynaptic spike are made less likely to contribute to the network.

Systems such as *ReSuMe* [9] or the *Tempotron* [10] used these in single-layer SNNs. Again, multilayer systems typically use the variants of backpropagation inherited from traditional NNs. Most multilayer or deep SNNs depend on some parameters that can be differentiated and on which backpropagation can be performed. A famous early system, *SpikeProp* [11] derives a backpropagation rule for spike times in the output layer. However, SpikeProp, because of the vast computational expense of backpropagation, has not been applied to large-scale problems.

Recently, Lee et al. [12] have used stochastic gradient descent on real-valued membrane potentials. Spike patterns are then subject to filtering before they are finally used for backpropagation. This method achieves state-of-the-art results for deep SNNs on tasks such as MNIST and N-MNIST. Recently, Kulkarni and Rajendran [13] used a similar technique and achieved similar results. They minimized the distance between network output and a regular firing target spike train for the desired output neuron.

Wu et al. [14] used a spatiotemporal backpropagation rule for SNN by separating spatial input signals, which come from other neurons, from temporal dynamics arising from the spiking behavior of the neuron itself. Jin et al. [15] used a backpropagation rule for a rate-coded error signal on a longer "macro-"timescale and updates on a shorter "micro-"timescale intended to capture individual spike effects.

However, a potential major problem exists with all of these systems; the error is usually computed at the output layer. Since the information flow in biological axons is assumed to be unidirectional, it is unclear how error information derived from learning methods descended from Hebbian or backpropagation can reach earlier parts of the neural network at all.

Traditionally, the answer to this problem is by utilizing bidirectional or random connections in order to provide recurrent feedback connections that modulate learning in earlier layers. This is the idea underlying echo state networks (ESN) or liquid state machines [16], and reservoir computing [17]. Random backpropagation has been postulated as one possible solution for training SNNs with backpropagation, by separating and synchronizing forward and backward passes, and by using symmetrical weights in both directions [18]. Research into energy-based networks and backpropagation [19] has also been performed. Recently Mozafari et al. [20] added multiple layers with reward-modulated spike time-dependent placidity (STDP) to a SNN to obtain fully spiking supervised training. In order to simultaneously train all layers with STDP within a deep network, Thiele et al. [21] introduced neurons with two integrate-and-fire units decoupling learning with STDP and inference. Autoencoders are trained layer-by-layer with unsupervised STDP [22].

However, the problem remains in that despite these modifications to the standard SNN model, there is no biological justification forms many of them. They are a pragmatic solution to a problem constrained by current technology and modus operandi. At its core are two questions:

- How to represent information in SNN, should weight changes constitute learning in such systems as they do in traditional neural networks?
- How should learning propagate in such a system?

1.1 The Weight Transport Problem and Credit Assignment

The weight transport problem has been around for almost as long as backpropagation itself [23].

It was recently reported that Geoffrey Hinton, the "father" of DL, said he was "deeply suspicious" of backpropagation, and said "my view is thrown it all away and start again."

He outlined a number of problems; for example, in real biological systems:

1. There is no obvious source for a supervision signal.
2. Cortical neurons do not communicate real-valued activities needed for backprop (they send all or none spikes).
3. Neurons do not have symmetrical reciprocal connections. The feedback connections do not go back to the neurons that the feedforward connections came from.
4. The neurons need to send two different types of signals (a forward pass: output = activity = y and a backward pass: output = error derivative w.r.t input = dE/dx).

However, Hinton mitigates the argument against backpropagation for each of the above problems by proposing how structures could be built that circumvent these neurological impediments.

Problems 1 and *2* can be easily addressed by using both self-regulatory systems and a different representation scheme for real values, such as used in spiking neural networks.

Problem 3 can be addressed by observing that the brain has many recurrent structures; these structures do not directly go back to the originating neuron but do often go back to the rough region of where the signal originated.

Problem 4 is a bit harder to address. At first, this seems in direct contradiction of one interpretation of Dale's law; sending two different signals simultaneously, in other words a single neuron representing two different quantities (error derivative and signal) depending on the different state of the system. However, Hinton, Bartunov, Bengio, and others have recently shown that by modifying the standard interpretation of STDP even this problem can be addressed; spike represents signals and the difference in spiking rates over some interval represents the error derivatives that are backpropagated in the network [24].

However, even if these problems can be addressed as Lillicrap [25] pointed out backpropagation is on the extreme end of a type of learning that may occur in the brain, and that there are undoubtedly others, but it is the only one that is being used in AI research at present. At the other end of the spectrum, learning could be done by global weight perturbation, but so far no learning mechanism has been found that is practicable has been found that can be used in even small-scale networks (let alone deep ones). Backpropagation can be viewed as an intensely precise instrument that requires significant time to repeatedly, synchronously, select neurons and update errors in the network, whereas global weight perturbation can be seen as a blunt instrument used to rapidly effect, asynchronously, global change on the whole network.

Between these two extremes, there is evidence of learning algorithms that are yet unexplored. Hinton recently examined capsules or groups of neurons whose outputs represent different properties of the same entity with each capsule operating on a similarity voting scheme and implement [26]. As early as 1995 Hinton attempted to use a "Wake-Sleep" algorithm, which used recognition weights in the wake phase and used generative weights in the sleep phase in order to explore different learning mechanisms on this continuum [27].

Recently, however, Bellec et al. [3] have come up with the idea of e-prop(agation) that enables network learning through gradient descent. This exciting development recognizes two key factors hitherto ignored in RSNN research:

(1) The dynamics of neurons in the brain is enriched by continuously ongoing updates of traces of past activity on the molecular level, for example, in the form of calcium ions.
(2) In the brain, there exists an abundance of top-down signals such as dopamine and acetylcholine, to name only a few, that inform local populations of neurons about sub-optimal performance of brain computations.

From this, they prove that traces and learning signals can be combined to produce network learning through gradient descent—without backpropagation of signals through time or by recurrent synaptic connections.

In this, model spikes are treated as binary variables (z_j^t) that is given a value of 1 if neuron j fires at time t. The goal of network learning is to find synaptic weight W that minimizes a given loss function E where E depends on all or a subset of the spikes in the network. E measures in the case of regression or classification learning the deviation of the actual output (y_k^t) of each output neuron k at time t from its given target value ($y_k^{*,t}$). In other words, E represents the effectiveness of a particular network of weights. In order to reduce E to a minimum, then the gradient $\frac{dE}{dW_{ji}}$ for the weight W_{ji} of the synapse from neuron i to neuron j indicates how it should be reduced. This gradient is seen as a sum over the time steps t of the RSNN computation.

While *e-propagation* correctly identifies the difficulties in the biological plausibility of *backpropagation* and also hints at calcium dynamics as being a possible solution, it is conjectured here that the interaction of calcium dynamics adds an extra level of interactivity that has the potential to dramatically increase the sophistication of the RSNN.

1.2 Representation of Weights in SNNs

Most modern SNNs use membrane potential, threshold function or, more even sometimes, spike delay to represent weights. The calcium control hypothesis says that a large calcium transient produces *long-term potentiation* (LTP), whereas a moderate increase in calcium results in *long-term depression* (LTD) [28]. This hypothesis has received significant experimental support in neuroscience.

To stimulate learning in such a system, it is necessary to define rules that translate calcium dynamics in the postsynaptic neuron to changes in synaptic strength.

Shouval [29] showed how this could be done:

$$\frac{dW_i}{dt} = \eta(\mathrm{Ca})(\Omega(\mathrm{Ca}) - \lambda \mathrm{Wi}) \quad (1)$$

Here W_i is the synaptic efficacy of synapse i, Ω determines the sign magnitude of synaptic plasticity as a function of the calcium (Ca) levels, η is a calcium-dependent learning rate, which is usually a monotonically increasing function of calcium, and λ is decay constant which can be generically set to $\lambda = 0$, no decay, or $\lambda = 1$. If $\lambda = 1$, for sustained calcium elevation, the synaptic efficacy converges to the value of Ω and the rate of convergence depends on η.

Learning takes place as neuron pairs become "*coincidence detectors*" by pre- and postsynaptic activity, in turn controlled by the blockage of the respective neuron's calcium receptors.

2 The Other Brain

However, none of the above considers lager-scale structures that may be present in the brain or how these "macro-"structures may be represented in a RSNN.

Whereas neurons provide fast point-to-point electrochemical signaling that may occur at large distances, glial cells work at a chemical only and a fairly local level. Neurons signals are fast (up to 268 mph); glial calcium signals are slow (Ca^{2+} wavefront velocity typically being 1.05 ± 0.17 μm/s).

Many neuroscientists are suggesting [30] that neurons and glia have a hitherto unsuspected complex interaction that forms the basis of cognition. While controversy currently exists about the extent and the exact mechanisms involved in calcium dynamics [31], there is a consensus that glia/neural interaction constitutes a complex electrochemical feedback mechanism. This is strongly suspected to be the principal agent in *long-term depression and pontentiation* (LTD and LTP) as both require activation of NMDA-type glutamate receptors and the resulting entry of Ca^{2+} into the postsynaptic cell. As such the process of learning is intrinsically linked to this activity. *Yet only a handful of artificial neural models try to reproduce this interaction.*

3 A Simple Model

We attempt to model calcium dynamics using the Belousov–Zhabotinsky (B–Z) reaction. Calcium levels can influence neural activity in a number of ways, such as the regulation of metabolic activity, the regulation of cell growth, and the long-term modification of synaptic efficiency and is also implicated in the destruction of neurons. It has been shown that a decrease in extracellular calcium levels brings about a reduction in the threshold for excitation of the nerve and may even induce spontaneous electrical activity [32]. Conversely, an increase in calcium causes reduced excitability. More significantly, for this work, it has been discovered that calcium transients can have a direct influence on synaptic plasticity and is therefore linked to Hebbian-like learning.

In our model, we use elevated calcium levels as a way to directly influence learning (in an effort to address the weight transport problem) and as a way to stimulate/depress spike activity. The interaction of these traveling *wavefronts*, similar to interference patterns on the ripple of a pond, is an ideal mechanism to distribute weight changes across the surface of the neural substrate.

The interaction of these calcium wavefronts provides a biologically plausible mechanism to strengthen and weaken connection in the network, and in other words, *wavefront interaction* is used as a contrivance for the network to learn. Furthermore, the mathematical principles behind the B–Z reaction are well understood and are a good fit for inclusion into a more complex spiking neural network simulation [33]. Not only does this obviate the biological implausibility of backpropagation, but it also has the potential to mimic, and expand upon, one of deep learning major breakthrough techniques, principally convolution.

Temporal convolution through the medium of calcium dynamics has the potential associate together groups of neurons activated at specific intervals of time. Different activation patterns would interfere with different groups of neurons. In this way, far more subtle patterns could be encoded onto the network, with each neuron playing a different role at different times. Such cooperation between the neurons in the network increases the potential to create a far more powerful classification network.

4 Proposed Systems

4.1 Diffusion Model

With this model, the calcium diffuses to the adjacent extracellular medium. Either a partial differential equation such as used in the heat equation or in Fick's first or second laws of diffusion can be used or a more complex model using continuous-time stochastic processes as used in Brownian motion can be used to model particle movement. In its simplest form, the standard diffusion equation can be used to calculate the Ca^{2+} flux from one cell to another; for example, the flux J is proportional to the gradient in concentration of Ca^{2+}, namely

$$J = D\frac{dn}{dx}$$

where D is the diffusion coefficient of the material and $\frac{dn}{dx}$ is the concentration of calcium between adjacent cells.

4.2 Wave Propagation Model

The wave propagation model is more biologically relevant than the simple diffusion model presented above. Glial networks generate calcium waves spreading across relatively large areas of the neural substrate. These waves interact to produce complex interference patterns. It is these interference patterns and the Ca^{2+} levels encoded within them that form the basis of this model. The wave model, unlike the diffusion model, recognizes that Ca^{2+} stores are not unlimited and must be used sparingly.

It has very recently been discovered that in real biological systems a rapidly oscillating calcium wave spread quickly but is much more localized than with a calcium wave that is slowly oscillating. This is probably due to the amplitude of the initial calcium wave being strongly influenced by the initial concentration of calcium released. The larger the release, the larger the amplitude of the wave, and the longer

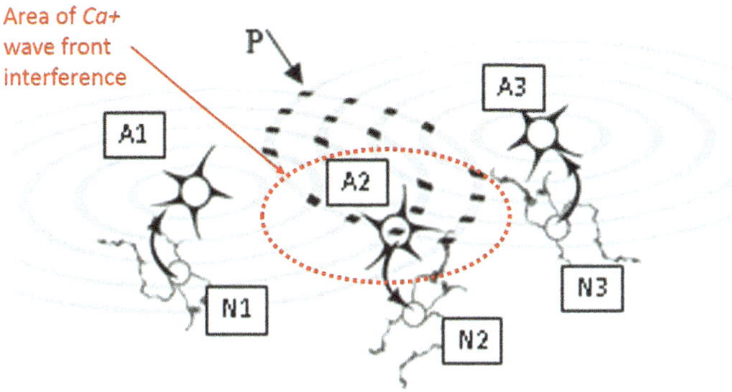

Fig. 1 Interaction of neuron and glial network

it will last. The smaller the release of calcium, the more quickly a new release can be made. Rapid oscillations will produce smaller amplitude and less longevity (in other words more suited to rapid localized interactions).

The proposed system uses the interactions between glia, synapse, and neuron to provide macroscopic function. The interference patterns of raised levels of Ca^{2+} waves modulate transmission properties of the surrounding synapses, which in turn has an effect on the functionality of the neurons they are connected to. This is illustrated in Fig. 1.

In Fig. 1, at time t_1, the neurons $N1$, $N2$, $N3$ are quiescent, as are astrocytes $A1$, $A2$, $A3$. At time $t_2 (t_2 > t_1)$, neurons $N1$ and $N3$ are responding to an input stimulus, in turn stimulating astrocytes $A1$ and $A2$. $A1$ and $A2$ emit calcium waves, which propagate through the surrounding medium. At time $t_3 (t_3 > t_2)$, points of constructive interference (P) will cause "hot spots" of glial activity (e.g., $A2$) stimulating nearby neurons (e.g., $N2$).

In this model, raised calcium levels to have *three major effects* on synaptic functionality:

1. It has the potential to vary the *postsynaptic potential* (PSP). For example, the model of the PSP of an excitatory or inhibitory synapse may, after raising calcium levels, have a much steeper initial climb and a much shallower tail. Moreover, the height and area under the curve are analogous to altering the size of the synaptic gap, and by implication, the ability for neurotransmitters to cross this gap. If raised calcium levels reduces the size of this gap, or increases the ability of the synapse to send or receive neurotransmitters, then the curve used to model the PSP changes. In other word, *synaptic plasticity*, or weight, changes. In this model, we use the equation below to simulate the effect of raised calcium levels. Where Q_{Sk} and V_{Sk} represent synaptic time constraints, the ratio between the two determines the initial steepness of the curve, the overall height of the curve, and the curve's decay (Fig. 2).

Fig. 2 Time patterns of the postsynaptic potential (PSP) with normal and elevated calcium level (top and bottom panels, respectively)

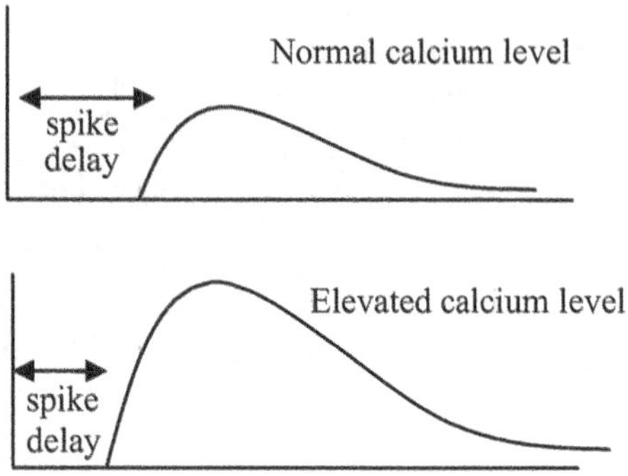

$$E(S_k) = \frac{S_k}{Q_{sk}}\left(1 - \boldsymbol{exp}\left(-\frac{S_k}{V_{sk}}\right)\right) \quad (2)$$

2. New synapses may be formed if sufficient dendritic/axonal activity occurs between neurons. This direction (and motility) of dendritic development may be coordinated by the direction and strength of the calcium waves. Moreover, the interference between waves tells the system about the (i) *direction of the waves*, (ii) *the location of the initial wave*, and (iii) *its immediate history*.

   The neuron itself is directly influenced by the synapses it is attached to and therefore indirectly affected by the calcium wave. Furthermore, the refractory period of the neuron may be decreased if the calcium levels around the neuron are raised. Again, this makes biological sense; if a neuron becomes exhausted, then the more stores of calcium it has to replenish itself, the shorter its refractory periods will be (Fig. 3).

3. The glial cell responds to neuronal activity through the elevation of intracellular Ca^{2+} levels. This causes the release of neurotransmitters which in turn

Fig. 3 Patterns of the refractory period with normal and elevated calcium levels

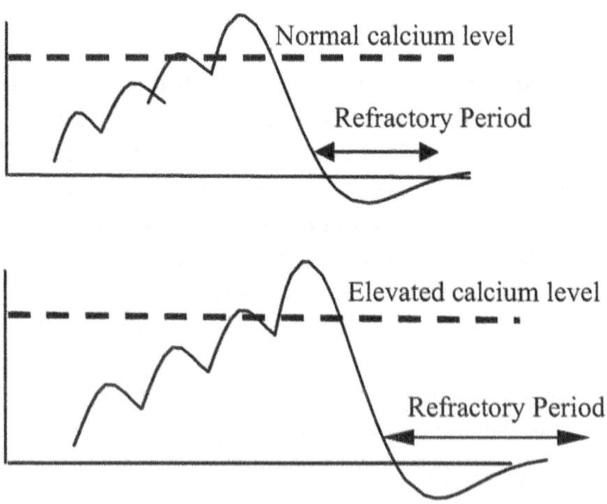

adapt the transmission of signals across the neural synapse, hence forming a feedback mechanism (which could be positive or negative depending on the type of neurotransmitter released).

5 Conclusion

This paper has shown how the glial network has biological relevance to RSNNs and proposes a proof-of-concept system. In particular, the paper has described how growing evidence for the central role of the glial network and calcium in neural stimulation/depression opens up the possibility of a step change in the computational power afforded by RSNN. More importantly, it gives biological justification for new categories of learning (such as q-prop) as an alternative to backpropagation type learning.

We believe that such a biological-inspired approach may benefit the development of a novel learning model as well as a better comprehension of the mechanism behind biological mechanism and, in particular, the learning process which can affect bio-mimetic and robotic applications.

References

1. M. Pfeiffer, T. Pfeil, Deep learning with spiking neurons: opportunities and challenges. Front. Neurosci. (Frontiers Media, 2018) **12**, Article 774
2. A. Tavanaei, M. Ghodrati, S.R. Kheradpisheh, T. Masquelier, A. Maida, Deep learning in spiking neural networks. Neural Netw. (Elsevier) **111**, 47–63 (2019)
3. G. Bellec, F. Scherr, A. Subramoney, E. Hajek, D. Salaj, R. Legenstein, W. Maass, A solution to the learning dilemma for recurrent networks of spiking neurons, bioRxiv (2019)
4. V. Demin, D. Nekhaev, Recurrent spiking neural network learning based on a competitive maximization of neuronal activity. Front. Neurosci. (Frontiers Media, 2018) **12**
5. R. Gütig, To spike, or when to spike?, in *Current Opinion in Neurobiology*, Theoretical and Computational Neuroscience Special Issue (Elsevier, 2014), pp. 134–139
6. M.A. Montemurro, M.J. Rasch, Y. Murayama, N.K. Logothetis, S. Panzeri, Phase-of-firing coding of natural visual stimuli in primary visual cortex. Curr. Biol. (Elsevier) **18**, 375–380 (2008)
7. W. Gerstner, R. Kempter, J.L. van Hemmen, H. Wagner, A neuronal learning rule for sub-millisecond temporal coding. Nature (Springer) **363**, 76–81 (1996)
8. F. Rieke, D. Warland, R. Steveninck, W. Bialek, in *Spikes: exploring the Neural Code* (MIT Press, 1999). ISBN:0-262-18174-6
9. F. Ponulak, A. Kasinski, Supervised learning in spiking neural networks with ReSuMe: sequence learning, classification, and spike shifting, in *Neural Computation* vol. 22 (MIT Press, 2010)
10. R. Gütig, H. Sompolinsky, The tempotron: a neuron that learns spike timing-based decisions, in *Nature Neuroscience*, vol. 9 (Springer, 2006)
11. S.M. Bohte, J.N. Kok, H. La Poutre, Spike-prop: error-backpropagation in multi-layer networks of spiking neurons. Neurocomputing (Elsevier) **48**, 17–37 (2002)

12. J.H. Lee, T. Delbruck, M. Pfeiffer, Training deep spiking neural networks using backpropagation. Front. Neurosci. (Frontiers Media, 2016) **10**
13. S.R. Kulkarni, B. Rajendran, Spiking neural networks for handwritten digit recognition-Supervised learning and network optimization. Neural Netw. (Elsevier) **10**, 118–127 (2018)
14. Y. Wu, L. Deng, G. Li, J. Zhu, L. Shi, Spatio-temporal backpropagation for training high-performance spiking neural networks. Front. Neurosci. (Frontiers Media) **12** (2018)
15. Y. Jin, W. Zhang, P. Li, Hybrid macro/micro level backpropagation for training deep spiking neural networks, in *Neural and Evolutionary Computing* (IEEE, 2019)
16. W. Maass, T. Natschläger, H. Markram, Real-time computing without stable states: a new framework for neural computation based on perturbations. Neural Comput (MIT Press) **14**, 2531–2560 (2002)
17. B. Schrauwen, D. Verstraeten, J. van Campenhout, An overview of reservoir computing: theory, applications, and implementations, in *Proceedings of the European Symposium on Artificial Neural Networks ESANN* (2007), pp. 471–482
18. Y. Bengio, D. Lee, J. Bornschein, T, Mesnard, Z. Lin, Towards biologically plausible deep learning (2015). arXiv:1502.04156
19. B. Scellier, Y. Bengio, Equilibrium propagation: bridging the gap between energy-based models and backpropagation. Front. Comput. Neurosci. (Frontiers Media), **11** (2017)
20. M. Mozafari, S.R. Kheradpisheh, T. Masquelier, A. Nowzari-Dalini, M. Ganjtabesh, First-spike-based visual categorization using reward-modulated STDP, in IEEE Trans. Neural Netw. Learn. Syst. (IEEE), **12** (2018)
21. J.C. Thiele, O. Bichler, A. Dupret, Event-based, timescale invariant unsupervised online deep learning with STDP. Front. Comput. Neurosci. (Frontiers Media), **12** (2018)
22. P. Panda, K. Roy, Unsupervised regenerative learning of hierarchical features in spiking deep networks for object recognition, in *Neural and Evolutionary Computing* (MIT Press, 2016)
23. S. Grossberg, Competitive learning: from interactive activation to adaptive resonance. Cogn. Sci. **11**(1), 23–63 (1987)
24. S. Bartunov, A. Santoro, B.A. Richards, L. Morris, G.E. Hinton, T.P. Lillicrap, Assessing the scalability of biologically-motivated deep learning algorithms and architectures, in *32nd Conference on Neural Information Processing Systems (NIPS)* (Montreal, Canada, 2018). arXiv:1807.04587v2
25. T.P. Lillicrap, Deep learning and the brain: does the brain approximate backpropagation?, in *Carnegie Mellon University's BrainHub Victor Bearg Lecture* (2018). https://www.youtube.com/watch?v=zQSNijL1fJg
26. G.E. Hinton, S. Sabour, N. Frosst, Matrix capsules with EM routing, in *Proceedings of the International Conference on Learning Representations (ICLR)* (Vancouver, Canada, 2018)
27. G.E. Hinton, P. Dayan, B.J. Frey, R. Neal, The wake-sleep algorithm for unsupervised neural networks. Sci. Am. Assoc. Adv. Sci. **268**(5214), 1158–1161 (1995)
28. J. Lisman, A mechanism for the Hebb and the anti-Hebb processes underlying learning and memory. Proceedings of the National Academy of Science United States of America **86**(23), 9574–9578 (1989)
29. H.Z. Shouval, G.C. Castellani, B.S. Blais, L.N. Cooper, Converging evidence for a simplified biophysical model of synaptic plasticity, in *Proceedings of Biological Cybernetics* (Springer, 2003)
30. G. Perea, M. Sur, A. Araque, Neuron-glia networks: integral gear of brain function. Front. Cell. Neurosci. (Frontiers Press) **8** (2014)
31. D.S. Auld, R. Robitaille, Glial cells and neurotransmission: an inclusive view of synaptic function. Neuron (Elsevier) **40**(2), 389–400 (2003)
32. B. Lu, Q. Zhang. H. Wang, Y. Wang, M. Nakayama, D. Ren, Extracellular calcium controls background current and neuronal excitability via an UNC79-UNC80-NALCN cation channel complex. Neuron (Elsevier), **68**(3), 488–499 (2010)
33. G. Chen, A mathematical model for bifurcations in a Belousov-Zhabotinsky reaction. Phys. D: Nonlinear Phenom. (Elsevier) **145**(3), 309–329 (2000)

Author Index

A
Abdulrahman, Rashid, 175
Agarwal, Shubham, 67
Agrawal, Suchitra, 127
Alam, Muhammad, 83
Assad, Assif, 141
Azeezunnisha, P.S., 57

B
Bansal, Jagdish Chand, 1
Biswas, Saroj Kr., 97
Buckley, Neil, 141, 199

C
Caddet, Philippe, 191
Chakraborty, Manomita, 97
Chaturvedi, Pratik, 67

D
Das, Dolly, 97
Das, Nipan, 97
Deep, Kusum, 141
Dutt, Varun, 67
Dwiyasa, Felis, 155, 167

F
Felmeri, Ian, 175
Foo, Ren-Xiang, 155, 167

G
Gedeon, Rawan, 47
Goel, Ishan, 127
Gupta, Ritu, 83, 111

H
Hemalatha, S., 57

I
Inyang, Udoinyang G., 27

J
Josman, Casey, 111

K
Kang, Pyo, 167
Kumar, Praveen, 67

L
Lim, Meng-Hiot, 155, 167

M
Mali, Naresh, 67

N
Nagar, Atulya Kumar, 47, 57, 141, 175, 191, 199
Naguib, Raouf, 47
Nyoho, Emmanuel E., 27

O
O'Connor, Declan, 199

© Springer Nature Singapore Pte Ltd. 2020

A. K. Nagar et al. (eds.), *Soft Computing for Problem Solving 2019*,
Advances in Intelligent Systems and Computing 1139,
https://doi.org/10.1007/978-981-15-3287-0

P
Pathania, Ankush, 67
Priyanka, 67
Purkayastha, Biswajit, 97

R
Reid, David, 211
Robertson, Sam, 111

S
Secco, Emanuele Lindo, 175, 191, 211
Sharma, Apoorva, 1
Sharma, Nirmala, 1
Sharma, Harish, 1
Shukla, Praveen Kumar, 17
Singh, Ravinder, 67
Sriram, Sastha, 57

T
Teo, Shi-Wei Jason, 155, 167
Tiwari, Aruna, 127
Tjia, Dewi, 83

U
Uday, K. V., 67
Umoh, Uduak A., 27
Upadhya, Rohit, 97

W
Wong, Lewis, 199

The manufacturer's authorised representative in the EU is Springer Nature Customer Service Centre GmbH, Europaplatz 3, 69115 Heidelberg, Germany. If you have any concerns regarding our products, please contact ProductSafety@springernature.com

Printed and bound by CPI Group (UK) Ltd, Croydon, CR0 4YY

30/09/2026

02225509-0006